Proofs That Really Count

The Art of Combinatorial Proof

© 2003 by

The Mathematical Association of America (Incorporated)

Library of Congress Catalog Card Number 2003108524

ISBN 10: 0-88385-333-7

ISBN 13: 978-0-88385-333-7

Printed in the United States of America

Current Printing (last digit):

10 9 8 7 6 5 4 3

The Dolciani Mathematical Expositions

NUMBER TWENTY-SEVEN

Proofs That Really Count

The Art of Combinatorial Proof

Arthur T. Benjamin
Harvey Mudd College
and
Jennifer J. Quinn
Occidental College

Published and Distributed by
The Mathematical Association of America

DOLCIANI MATHEMATICAL EXPOSITIONS

Committee on Publications
GERALD ALEXANDERSON, *Chair*

The DOLCIANI MATHEMATICAL EXPOSITIONS series of the Mathematical Association of America was established through a generous gift to the Association from Mary P. Dolciani, Professor of Mathematics at Hunter College of the City University of New York. In making the gift, Professor Dolciani, herself an exceptionally talented and successful expositor of mathematics, had the purpose of furthering the ideal of excellence in mathematical exposition.

The Association, for its part, was delighted to accept the gracious gesture initiating the revolving fund for this series from one who has served the Association with distinction, both as a member of the Committee on Publications and as a member of the Board of Governors. It was with genuine pleasure that the Board chose to name the series in her honor.

The books in the series are selected for their lucid expository style and stimulating mathematical content. Typically, they contain an ample supply of exercises, many with accompanying solutions. They are intended to be sufficiently elementary for the undergraduate and even the mathematically inclined high-school student to understand and enjoy, but also to be interesting and sometimes challenging to the more advanced mathematician.

1. *Mathematical Gems,* Ross Honsberger

2. *Mathematical Gems II,* Ross Honsberger

3. *Mathematical Morsels,* Ross Honsberger

4. *Mathematical Plums,* Ross Honsberger (ed.)

5. *Great Moments in Mathematics (Before 1650),* Howard Eves

6. *Maxima and Minima without Calculus,* Ivan Niven

7. *Great Moments in Mathematics (After 1650),* Howard Eves

8. *Map Coloring, Polyhedra, and the Four-Color Problem,* David Barnette

9. *Mathematical Gems III,* Ross Honsberger

10. *More Mathematical Morsels,* Ross Honsberger

11. *Old and New Unsolved Problems in Plane Geometry and Number Theory,* Victor Klee and Stan Wagon

12. *Problems for Mathematicians, Young and Old,* Paul R. Halmos

13. *Excursions in Calculus: An Interplay of the Continuous and the Discrete,* Robert M. Young

14. *The Wohascum County Problem Book,* George T. Gilbert, Mark Krusemeyer, and Loren C. Larson

15. *Lion Hunting and Other Mathematical Pursuits: A Collection of Mathematics, Verse, and Stories by Ralph P. Boas, Jr.,* edited by Gerald L. Alexanderson and Dale H. Mugler

16. *Linear Algebra Problem Book,* Paul R. Halmos

17. *From Erdös to Kiev: Problems of Olympiad Caliber,* Ross Honsberger

18. *Which Way Did the Bicycle Go? ...and Other Intriguing Mathematical Mysteries,* Joseph D. E. Konhauser, Dan Velleman, and Stan Wagon

19. *In Pólya's Footsteps: Miscellaneous Problems and Essays,* Ross Honsberger

20. *Diophantus and Diophantine Equations,* I. G. Bashmakova (Updated by Joseph Silverman and translated by Abe Shenitzer)

21. *Logic as Algebra,* Paul Halmos and Steven Givant

22. *Euler: The Master of Us All,* William Dunham

23. *The Beginnings and Evolution of Algebra,* I. G. Bashmakova and G. S. Smirnova (Translated by Abe Shenitzer)

24. *Mathematical Chestnuts from Around the World,* Ross Honsberger

MAA Service Center
P.O. Box 91112
Washington, DC 20090-1112
1-800-331-1MAA   FAX: 1-301-206-9789

Dedicated to our families—the people who count the most in our lives.

To Deena, Laurel, and Ariel. A.T.B.

To the bad boys of Loleta. J.J.Q.

Foreword

Every proof in this book is ultimately reduced to a counting problem—typically enumerated in two different ways. Counting leads to beautiful, often elementary, and very concrete proofs. While not necessarily the simplest approach, it offers another method to gain understanding of mathematical truths. To a combinatorialist, this kind of proof is the *only* right one. We offer *Proofs That Really Count* as the counting equivalent of the visual approach taken by Roger Nelsen in *Proofs Without Words I & II* [37, 38].

Why count?

As human beings we learn to count from a very early age. A typical 2 year old will proudly count to 10 for the coos and applause of adoring parents. Though many adults readily claim ineptitude in mathematics, no one ever owns up to an inability to count. Counting is one of our first tools, and it is time to appreciate its full mathematical power. The physicist Ernst Mach even went so far as to say, "There is no problem in all mathematics that cannot be solved by direct counting" [36].

Combinatorial proofs can be particularly powerful. To this day, I (A.T.B.) remember my first exposure to combinatorial proof when I was a freshman in college. My professor proved the Binomial Theorem

$$(x + y)^n = \sum_{k=0}^{n} \binom{n}{k} x^k y^{n-k}$$

by writing

$$(x + y)^n = \underbrace{(x + y)(x + y) \cdots (x + y)}_{n \text{ times}}$$

and asking "In how many ways can we create an $x^k y^{n-k}$ term?" Sudden clarity ensued. The theorem made perfect sense. Yes, I had seen proofs of the Binomial Theorem before, but they had seemed awkward and I wondered how anyone in his or her right mind would create such a result. But now it seemed very natural. It became a result I would never forget.

What to count?

We have selected our favorite identities using numbers that arise frequently in mathematics (binomial coefficients, Fibonacci numbers, Stirling numbers, etc.) and have chosen elegant counting proofs. In a typical identity, we pose a counting question, and then answer it in two different ways. One answer is the left side of the identity; the other answer is the right side. Since both answers solve the same counting question, they must be equal. Thus the identity can be viewed as a counting problem to be tackled from two different angles.

We use the identity

$$\sum_{k=0}^{n} \binom{n}{k} = 2^n$$

to illustrate a proof structure found throughout this book. There is no need to use the formula $\frac{n!}{k!(n-k)!}$ for $\binom{n}{k}$. Instead, we interpret $\binom{n}{k}$ as the number of k-element subsets of an n-element set, or more colorfully, as the number of ways to select a committee of k students from a class of n students.

Question: From a class of n students, how many ways can we create a committee?

Answer 1: The number of committees with 0 students is $\binom{n}{0}$. The number of committees with 1 student is $\binom{n}{1}$. In general, the number of committees with exactly k students is $\binom{n}{k}$. Hence the total number of committees is $\sum_{k=0}^{n} \binom{n}{k}$.

Answer 2: To create a committee of arbitrary size, we decide, student by student whether or not they will be on the committee. Since each of the n students is either "on" or "off" the committee, there are 2 possibilities for each student and thus 2^n ways to create a committee.

Since our logic is impeccable in both answers, they must be equal, and the identity follows.

Another useful proof technique is to interpret the left side of an identity as the size of a set, the right side of the identity as the size of a different set, and then find a one-to-one correspondence between the two sets. We illustrate this proof structure with the identity

$$\sum_{k\geq 0} \binom{n}{2k} = \sum_{k\geq 0} \binom{n}{2k+1} \quad \text{for } n > 0.$$

Both sums are finite since $\binom{n}{i} = 0$ whenever $i > n$. Here it is easy to see *what* both sides count. The challenge is to find the correspondence between them.

Set 1: The committees with an even number of members formed from a class of n students. This set has size $\sum_{k\geq 0} \binom{n}{2k}$.

Set 2: The committees with an odd number of members formed from a class of n students. This set has size $\sum_{k\geq 0} \binom{n}{2k+1}$.

Correspondence: Suppose one of the students in the class is named Waldo. Any committee with an even number of members can be turned into a committee with an odd number of members by asking "Where's Waldo?" If Waldo is on the committee, then remove him. If Waldo is not on the committee, then add him. Either way, the parity of the committee has changed from even to odd.

Since the process of "removing or adding Waldo" is completely reversible, we have a one-to-one correspondence between these sets. Thus both sets must have the same size, and the identity follows.

Often we shall prove an identity more than one way, if we think a second proof can bring new insight to the problem. For instance, the last identity can be handled by counting the number of even subsets directly. See Identity 129 and the subsequent discussion.

What can you expect when reading this book? Chapter 1 introduces a combinatorial interpretation of Fibonacci numbers as square and domino tilings, which serves as the foundation for Chapters 2–4. We begin here because Fibonacci numbers are intrinsically interesting and their interpretation as combinatorial objects will come as a delightful surprise to many readers. As with all the chapters, this one begins with elementary identities and simple arguments that help the reader to gain a familiarity with the concepts before proceeding to more complex material. Expanding on the Fibonacci tilings will enable us to explore identities involving generalized Fibonacci numbers including Lucas numbers (Chapter 2), arbitrary linear recurrences (Chapter 3), and continued fractions (Chapter 4.)

Chapter 5 approaches the traditional combinatorial subject of binomial coefficients. Counting sets with and without repetition leads to identities involving binomial coefficients. Chapter 6 looks at binomial identities with alternating signs. By finding correspondences between sets with even numbers of elements and sets with odd numbers of elements, we avoid using the familiar method of overcounting and undercounting provided by the Principle of Inclusion-Exclusion.

Harmonic numbers, like continued fractions, are not integral—so a combinatorial explanation requires investigating the numerator and denominator of a particular representation. Harmonic numbers are connected to Stirling numbers of the first kind. Chapter 7 investigates and exploits this connection in addition to identities involving Stirling numbers of the second kind.

Chapter 8 considers more classical results from arithmetic, number theory, and algebra including the sum of consecutive integers, the sum of consecutive squares, sum of consecutive cubes, Fermat's Little Theorem, Wilson's Theorem, and a partial converse to Lagrange's Theorem.

In Chapter 9, we tackle even more complex Fibonacci and binomial identities. These identities require ingenious arguments, the introduction of colored tiles, or probabilistic models. They are perhaps the most challenging in the book, but well worth your time.

Occasionally, we digress from identities to prove fun applications. Look for a divisibility proof on Fibonacci numbers in Chapter 1, a magic trick in Chapter 2, a shortcut to calculate the parity of binomial coefficients in Chapter 5 and generalizations to congruences modulo arbitrary primes in Chapter 8.

Each chapter, except the last, includes a set of exercises for the enthusiastic reader to try his or her own counting skills. Most chapters contain a list of identities for which combinatorial proofs are still being sought. Hints and references for the exercises and a complete listing of all the identities can be found in the appendices at the end of the book.

Our hope is that each chapter can stand independently, so that you can read in a nonlinear fashion if desired.

Who should count?

The short answer to this question is "Everybody counts!" We hope this book can be enjoyed by readers without special training in mathematics. Most of the proofs in this book can be appreciated by students at the high school level. On the other hand, teachers may find this book to be a valuable resource for classes that emphasize proof writing and creative problem solving techniques. We do not consider this book to be a complete survey of combinatorial proofs. Rather, it is a beginning. After reading it, you will never view quantities like Fibonacci numbers and continued fractions the same way again. Our hope is that an identity like

Identity 5.
$$f_{2n+1} = \sum_{i=0}^{n} \sum_{j=0}^{n} \binom{n-i}{j} \binom{n-j}{i}$$

for Fibonacci numbers should give you the feeling that something is being counted and the desire to count it. Finally, we hope this book will serve as an inspiration for mathematicians who wish to discover combinatorial explanations for old identities or discover new ones. We invite you, our readers, to share your favorite combinatorial proofs with us for (possible) future editions.

After all, we hope all of our efforts in writing this book will count for something.

Who counts?

We are pleased to acknowledge the many people who made this book possible—either directly or indirectly.

Those who came before us are responsible for the rise in popularity of combinatorial proof. Books whose importance cannot be overlooked are *Constructive Combinatorics* by Dennis Stanton and Dennis White, *Enumerative Combinatorics Volumes 1 & 2* by Richard Stanley, *Combinatorial Enumeration* by Ian Goulden and David Jackson, and *Concrete Mathematics* by Ron Graham, Don Knuth & Oren Patashnik. In addition to these mathematicians, others whose works continue to inspire us include George E. Andrews, David Bressoud, Richard Brualdi, Leonard Carlitz, Ira Gessel, Adriano Garsia, Ralph Grimaldi, Richard Guy, Stephen Milne, Jim Propp, Marta Sved, Herbert Wilf, and Doron Zeilberger.

One of the benefits of seeking combinatorial proofs is being able to involve undergraduate researchers. Many thanks to Robin Baur, Tim Carnes, Dan Cicio, Karl Mahlburg, Greg Preston, and especially Chris Hanusa, David Gaebler, Robert Gaebler, and Jeremy Rouse, who were supported through undergraduate research grants provided by the Harvey Mudd College Beckman Research Fund, the Howard Hughes Medical Institute, and the Reed Institute for Decision Science directed by Janet Myhre. Colleagues providing ideas, identities, input, or invaluable information include Peter G. Anderson, Bob Beals, Jay Cordes, Duane DeTemple, Persi Diaconis, Ira Gessel, Tom Halverson, Melvin Hochster, Dan Kalman, Greg Levin, T.S. Michael, Mike Orrison, Rob Pratt, Jim Propp, James Tanton, Doug West, Bill Zwicker, and especially Francis Su. It couldn't have happened without the encouragement of Don Albers and the work of Dan Velleman and the Dolciani board of the Mathematical Association of America. Finally, we are ever grateful for the love and support of our families.

Contents

CHAPTER 1

Fibonacci Identities

Definition The *Fibonacci numbers* are defined by $F_0 = 0$, $F_1 = 1$, and for $n \geq 2$, $F_n = F_{n-1} + F_{n-2}$.

The first few numbers in the sequence of Fibonacci numbers are 0, 1, 1, 2, 3, 5, 8, 13, 21, 34, 55, 89, 144, ...

1.1 Combinatorial Interpretation of Fibonacci Numbers

How many sequences of 1s and 2s sum to n? Let's call the answer to this counting question f_n. For example, $f_4 = 5$ since 4 can be created in the following 5 ways:

$$1 + 1 + 1 + 1, \quad 1 + 1 + 2, \quad 1 + 2 + 1, \quad 2 + 1 + 1, \quad 2 + 2.$$

Table 1.1 illustrates the values of f_n for small n. The pattern is unmistakable; f_n begins like the Fibonacci numbers. In fact, f_n will continue to grow like Fibonacci numbers, that is for $n > 2$, f_n satisfies $f_n = f_{n-1} + f_{n-2}$. To see this combinatorially, we consider the first number in our sequence. If the first number is 1, the rest of the sequence sums to $n - 1$, so there are f_{n-1} ways to complete the sequence. If the first number is 2, there are f_{n-2} ways to complete the sequence. Hence, $f_n = f_{n-1} + f_{n-2}$.

For our purposes, we prefer a more visual representation of f_n. By thinking of the 1s as representing *squares* and the 2s as representing *dominoes*, f_n counts the number of ways to *tile* a board of length n with squares and dominoes. For simplicity, we call a length n board an *n-board*. Thus $f_4 = 5$ enumerates the tilings:

Figure 1.1. All five square-domino tilings of the 4-board

We let $f_0 = 1$ count the empty tiling of the 0-board and define $f_{-1} = 0$. This leads to a combinatorial interpretation of the Fibonacci numbers.

Combinatorial Theorem 1 *Let f_n count the ways to tile a length n board with squares and dominoes. Then f_n is a Fibonacci number. Specifically, for $n \geq -1$,*

$$f_n = F_{n+1}.$$

1	2	3	4	5	6
1	11	111	1111	11111	111111
	2	12	112	1112	11112
		21	121	1121	11121
			211	1211	11211
			22	122	1122
				2111	12111
				212	1212
				221	1221
					21111
					2112
					2121
					2211
					222
$f_1 = 1$	$f_2 = 2$	$f_3 = 3$	$f_4 = 5$	$f_5 = 8$	$f_6 = 13$

Table 1.1. f_n and the sequence of 1s and 2s summing to n for $n = 1, 2, \ldots, 6$.

1.2   Identities

Elementary Identities

Mathematics is the science of patterns. As we shall see, the Fibonacci numbers exhibit many beautiful and surprising relationships. Although Fibonacci identities can be proved by a myriad of methods, we find the combinatorial approach ultimately satisfying.

For combinatorial convenience, we shall express most of our identities in terms of f_n instead of F_n. Although other combinatorial interpretations of Fibonacci numbers exist (see exercises 1–9), we shall primarily use the tiling definition given here.

In the proof of our first identity, as with most proofs in this book, one of the answers to the counting question breaks the problem into disjoint cases depending on some property. We refer to this as *conditioning* on that property.

Identity 1 *For $n \geq 0$, $f_0 + f_1 + f_2 + \cdots + f_n = f_{n+2} - 1$.*

Question: How many tilings of an $(n + 2)$-board use at least one domino?

Answer 1: There are f_{n+2} tilings of an $(n + 2)$-board. Excluding the "all square" tiling gives $f_{n+2} - 1$ tilings with at least one domino.

Answer 2: Condition on the location of the last domino. There are f_k tilings where the last domino covers cells $k + 1$ and $k + 2$. This is because cells 1 through k can be tiled in f_k ways, cells $k + 1$ and $k + 2$ must be covered by a domino, and cells $k + 3$ through $n + 2$ must be covered by squares. Hence the total number of tilings with at least one domino is $f_0 + f_1 + f_2 + \cdots + f_n$ (or equivalently $\sum_{k=0}^{n} f_k$). See Figure 1.2.

Identity 2 *For $n \geq 0$, $f_0 + f_2 + f_4 + \cdots + f_{2n} = f_{2n+1}$.*

Question: How many tilings of a $(2n + 1)$-board exist?

Answer 1: By definition, there are f_{2n+1} such tilings.

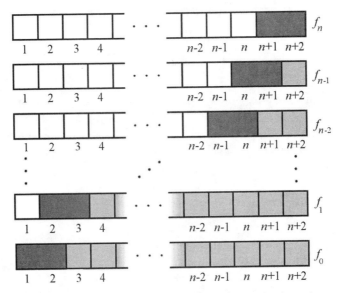

Figure 1.2. To see that $f_0 + f_1 + f_2 + \cdots + f_n = f_{n+2} - 1$, tile an $(n+2)$-board with squares and dominoes and condition on the location of the last domino.

Answer 2: Condition on the location of the last square. Since the board has odd length, there must be at least one square and the last square occupies an odd-numbered cell. There are f_{2k} tilings where the last square occupies cell $2k + 1$, as illustrated in Figure 1.3. Hence the total number of tilings is $\sum_{k=0}^{n} f_{2k}$.

Many Fibonacci identities depend on the notion of breakability at a given cell. We say that a tiling of an n-board is *breakable* at cell k, if the tiling can be decomposed into two tilings, one covering cells 1 through k and the other covering cells $k+1$ through n. On the

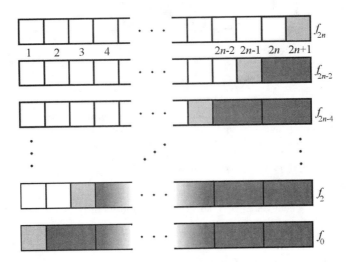

Figure 1.3. To see that $f_0 + f_2 + f_4 + \cdots + f_{2n} = f_{2n+1}$, tile a $(2n + 1)$-board with squares and dominoes and condition on the location of the last square.

Figure 1.4. A 10-tiling that is breakable at cells $1, 2, 3, 5, 7, 8, 10$ and unbreakable at cells $4, 6, 9$.

other hand, we call a tiling *unbreakable* at cell k if a domino occupies cells k and $k + 1$. For example, the tiling of the 10-board in Figure 1.4 is breakable at cells $1, 2, 3, 5, 7, 8, 10$, and unbreakable at cells $4, 6, 9$. Notice that a tiling of an n-board (henceforth abbreviated an *n-tiling*) is always breakable at cell n. We apply these ideas to the next identity.

Identity 3 *For $m, n \geq 0$, $f_{m+n} = f_m f_n + f_{m-1} f_{n-1}$.*

  Question: How many tilings of an $(m + n)$-board exist?

  Answer 1: There are f_{m+n} $(m + n)$-tilings.

  Answer 2: Condition on breakability at cell m.

  An $(m + n)$-tiling that is breakable at cell m, is created from an m-tiling followed by an n-tiling. There are $f_m f_n$ of these.

  An $(m + n)$-tiling that is unbreakable at cell m must contain a domino covering cells m and $m + 1$. So the tiling is created from an $(m - 1)$-tiling followed by a domino followed by an $(n - 1)$-tiling. There are $f_{m-1} f_{n-1}$ of these.

  Since a tiling is either breakable or unbreakable at cell m, there are $f_m f_n + f_{m-1} f_{n-1}$ tilings altogether. See Figure 1.5.

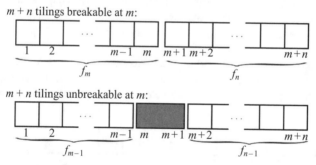

Figure 1.5. To prove $f_{m+n} = f_m f_n + f_{m-1} f_{n-1}$ count $(m + n)$-tilings based on whether or not they are breakable or unbreakable at m.

The next two identities relate Fibonacci numbers to binomial coefficients. We shall say more about combinatorial proofs with binomial coefficients in Chapter 5. For now, recall the following combinatorial definition for binomial coefficients.

Definition The *binomial coefficient* $\binom{n}{k}$ is the number of ways to select k elements from an n-element set.

Notice that $\binom{n}{k} = 0$ whenever $k > n$, so the sum in the identity below is finite.

Identity 4 *For $n \geq 0$, $\binom{n}{0} + \binom{n-1}{1} + \binom{n-2}{2} + \cdots = f_n$.*

  Question: How many tilings of an n-board exist?

  Answer 1: There are f_n n-tilings.

Answer 2: Condition on the number of dominoes. How many n-tilings use exactly i dominoes? For the answer to be nonzero, we must have $0 \leq i \leq n/2$. Such tilings necessarily use $n - 2i$ squares and therefore use a total of $n - i$ tiles. For example, Figure 1.6 is a 10-tiling that uses exactly three dominoes and four squares. The dominoes occur as the fourth, fifth, and seventh tiles. The number of ways to select i of these $n - i$ tiles to be dominoes is $\binom{n-i}{i}$. Hence there are $\sum_{i \geq 0} \binom{n-i}{i}$ n-tilings.

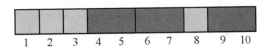

Figure 1.6. There are $\binom{7}{3}$ 10-tilings that use exactly three dominoes. Such a 10-tiling uses exactly seven tiles and is defined by which three of the seven tiles are dominoes. Here the fourth, fifth, and seventh tiles are dominoes.

Identity 5 *For* $n \geq 0$, $\displaystyle\sum_{i \geq 0}\sum_{j \geq 0} \binom{n-i}{j}\binom{n-j}{i} = f_{2n+1}.$

Question: How many tilings of a $(2n + 1)$-board exist?

Answer 1: There are f_{2n+1} $(2n + 1)$-tilings.

Answer 2: Condition on the number of dominoes on each side of the *median* square.

Any tiling of a $(2n+1)$-board must contain an odd number of squares. Thus one square, which we call the median square, contains an equal number of squares to the left and right of it. For example, the 13-tiling in Figure 1.7 has five squares. The median square, the third square, is located in cell 9.

How many tilings contain exactly i dominoes to the left of the median square and exactly j dominoes to the right of the median square? Such a tiling has $(i + j)$ dominoes and therefore $(2n + 1) - 2(i + j)$ squares. Hence the median square has $n - i - j$ squares on each side of it. Since the left side has $(n - i - j) + i = n - j$ tiles, of which i are dominoes, there are $\binom{n-j}{i}$ ways to tile to the left of the median square. Similarly, there are $\binom{n-i}{j}$ ways to tile to the right of the median square. Hence there are $\binom{n-i}{j}\binom{n-j}{i}$ tilings altogether.

As i and j vary, we obtain the total number of $(2n + 1)$-tilings as

$$\sum_{i \geq 0}\sum_{j \geq 0} \binom{n-i}{j}\binom{n-j}{i}.$$

median square

Figure 1.7. The 13-tiling above has three dominoes left of the median square and one domino to the right of the median square. The number of such tilings is $\binom{5}{3}\binom{3}{1}$.

The next identity is 'prettier' when stated as $F_{2n} = \sum_{k=0}^{n}\binom{n}{k}F_k$.

Identity 6 *For $n \geq 0$, $f_{2n-1} = \sum_{k=1}^{n} \binom{n}{k} f_{k-1}$.*

Question: How many $(2n - 1)$-tilings exist?

Answer 1: f_{2n-1}.

Answer 2: Condition on the number of squares that appear among the first n tiles. Observe that a $(2n - 1)$-tiling must include at least n tiles, of which at least one is a square. If the first n tiles consist of k squares and $n - k$ dominoes, then these tiles can be arranged $\binom{n}{k}$ ways and cover cells 1 through $2n - k$. The remaining board has length $k - 1$ and can be tiled f_{k-1} ways. See Figure 1.8.

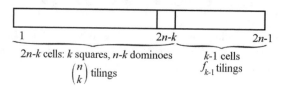

Figure 1.8. There are $\binom{n}{k} f_{k-1}$ tilings of a $(2n-1)$-board where the first n tiles contain k squares and $n - k$ dominoes.

For the next identity, we use the combinatorial technique of finding a correspondence between two sets of objects. In particular, we use a 1-to-3 correspondence between the set of n-tilings and the set of $(n - 2)$-tilings and $(n + 2)$-tilings.

Identity 7 *For $n \geq 1$, $3f_n = f_{n+2} + f_{n-2}$.*

Set 1: Tilings of an n-board. By definition, this set has size f_n.

Set 2: Tilings of an $(n+2)$-board or an $(n-2)$-board. This set has size $f_{n+2} + f_{n-2}$.

Correspondence: To prove the identity, we establish a *1-to-3 correspondence* between Set 1 and Set 2. That is, for every object in Set 1, we can create three objects in Set 2 in such a way that every object in Set 2 is created exactly once. Hence Set 2 is three times as large as Set 1.

Specifically, for each n-tiling in Set 1, we create the following three tilings that have length $n + 2$ or length $n - 2$. The first tiling is an $(n + 2)$-tiling created by appending a domino to the n-tiling. The second tiling is an $(n + 2)$-tiling created by appending two squares to the n-tiling. So far, so good. But what about the third tiling? This will depend on the last tile of the n-tiling. If the n-tiling ends with a square, we insert a domino before that last square to create an $(n + 2)$-tiling. If the n-tiling ends with a domino, then we remove that domino to create an $(n - 2)$-tiling. See Figure 1.9.

To verify that this is a 1-to-3 correspondence, one should check that every tiling of length $n + 2$ or length $n - 2$ is created exactly once from some n-tiling. For a given $(n + 2)$-tiling, we can find the n-tiling that creates it by examining its ending and removing

  i) the last domino (if it ends with a domino) or
 ii) the last two squares (if it ends with two squares) or
iii) the last domino (if it ends with a square preceded by a domino).

For a given $(n - 2)$-tiling, we simply append a domino for the n-tiling that creates it.

Since Set 2 is three times the size of Set 1, it follows that $f_{n+2} + f_{n-2} = 3f_n$.

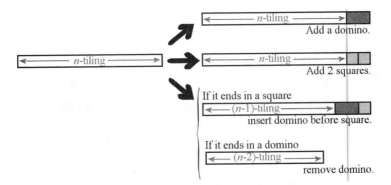

Figure 1.9. A one-to-three correspondence.

Pairs of Tilings

In this subsection, we introduce the technique of *tail swapping*, which will prove to be very useful in several settings.

Consider the two 10-tilings offset as in Figure 1.10. The first one tiles cells 1 through 10; the second one tiles cells 2 through 11. We say that there is a *fault* at cell i, for $2 \leq i \leq 10$, if both tilings are breakable at cell i. We say there is a fault at cell 1 if the first tiling is breakable at cell 1. Put another way, the pair of tilings has a fault at cell i, for $1 \leq i \leq 10$, if neither tiling has a domino covering cells i and $i + 1$. The pair of tilings in Figure 1.10 has faults at cells 1, 2, 5, and 7. We define the *tails* of a tiling pair to be the tiles that occur after the last fault. Observe that if we swap the tails of Figure 1.10 we obtain the 11-tiling and the 9-tiling in Figure 1.11, and it has the same faults.

Tail swapping is the basis for the identity below, sometimes referred to as Simson's Formula or Cassini's Identity. At first glance, it may appear unsuitable for combinatorial proof due to the presence of the $(-1)^n$ term. Nonetheless, we will see that this term is merely the "error term" of an "almost" one-to-one correspondence.

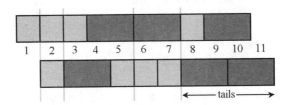

Figure 1.10. Two 10-tilings with their faults (indicated with gray lines) and tails.

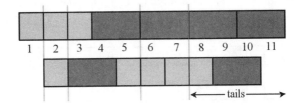

Figure 1.11. After tail swapping, we have an 11-tiling and a 9-tiling with exactly the same faults.

Identity 8 *For $n \geq 0$, $f_n^2 = f_{n+1}f_{n-1} + (-1)^n$*

Set 1: Tilings of two n-boards (a *top* board and a *bottom* board.) By definition, this set has size f_n^2.

Set 2: Tilings of an $(n+1)$-board and an $(n-1)$-board. This set has size $f_{n+1}f_{n-1}$.

Correspondence: First, suppose n is odd. Then the top and bottom board must each have at least one square. Notice that a square in cell i of either board ensures that a fault must occur at cell i or cell $i-1$. Swapping the tails of the two n-tilings produces an $(n+1)$-tiling and an $(n-1)$-tiling with the same faults. This produces a 1-to-1 correspondence between all pairs of n tilings and all tiling pairs of sizes $n+1$ and $n-1$ that have faults. Is it possible for a tiling pair of sizes $n+1$ and $n-1$ to be "fault-free"? Yes, precisely when all dominoes are in "staggered formation" as in Figure 1.12. Thus, when n is odd, $f_n^2 = f_{n+1}f_{n-1} - 1$.

Similarly, when n is even, tail swapping creates a 1-to-1 correspondence between faulty tiling pairs. The only fault-free tiling pair is the all domino tiling of Figure 1.13. Hence when n is even, $f_n^2 = f_{n+1}f_{n-1} + 1$. Considering the odd and even case together produces our identity.

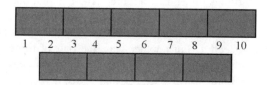

Figure 1.12. When n is odd, there is only one fault-free tiling pair.

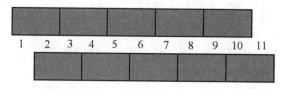

Figure 1.13. When n is even, there is only one fault-free tiling pair.

Identity 9 *For $n \geq 0$, $\sum_{k=0}^{n} f_k^2 = f_n f_{n+1}$.*

Question: How many tilings of an n-board and $(n+1)$-board exist?

Answer 1: There are $f_n f_{n+1}$ such tilings.

Answer 2: Place the $(n+1)$-board directly above the n board as in Figure 1.14, and condition on the location of the last fault. Since both boards begin at cell 1, we

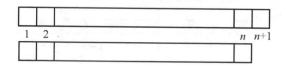

Figure 1.14. There are $f_n f_{n+1}$ ways to tile these two boards.

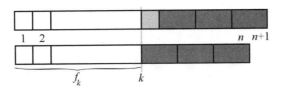

Figure 1.15. There are f_k^2 tilings with last fault at cell k.

shall consider any tiling pair to have a fault at "cell 0". How many tiling pairs have their last fault at cell k, where $0 \le k \le n$? There are f_k^2 ways to tile both boards through cell k. To avoid future faults, there is exactly one way to finish the tiling, as in Figure 1.15. (Specifically, all tiles after cell k will be dominoes except for a single square placed on cell $k+1$ in the row whose tail length is odd.) Summing over all possible values of k, gives us $\sum_{k=0}^{n} f_k^2$ tilings.

Advanced Fibonacci Identities

In this subsection we present identities that in our opinion require extra ingenuity. For the first identity, we utilize a method of encoding tilings as binary sequences.

Specifically, for any m-tiling, create the length m binary sequence by converting each square into a "1" and converting each domino into a "01". Equivalently, the ith term of the binary sequence is 1 if and only if the tiling is breakable at cell i. The resulting binary sequence will have no consecutive 0s and will always end with 1. For example, the 9-tiling in Figure 1.16 has binary representation 011101011.

Figure 1.16. The 9-tiling above has binary representation 011101011.

Conversely, a length n binary sequence with no consecutive 0s that ends with 1 represents a unique n-tiling. If such a sequence ends with 0, then it represents an $(n-1)$-tiling (since the last 0 is ignored).

We may now interpret the following identity.

Identity 10 *For $n \ge 0$,* $f_n + f_{n-1} + \sum_{k=0}^{n-2} f_k 2^{n-2-k} = 2^n$.

Question: How many binary sequences of length n exist?

Answer 1: There are 2^n length n binary sequences.

Answer 2: For each binary sequence, we identify a tiling. If a sequence has no consecutive zeros, we identify it with a unique tiling of length n or $n-1$ depending on whether it ended with 1 or 0, respectively. Otherwise, the sequence contains a 00 whose first occurrence appears in cells $k+1$ and $k+2$ for some k, $0 \le k \le n-2$. For such a sequence we associate the k-tiling defined by the first k terms of the binary sequence (note that if $k > 0$, then the kth digit must be 1.) For example, the length 11 binary sequence 01101001001 is identified with the 5-tiling "domino-square-domino", as would any binary sequence of the form 0110100$abcd$ where a, b, c, d

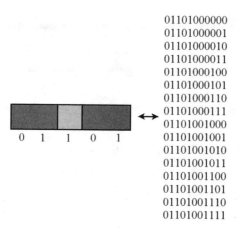

01101000000
01101000001
01101000010
01101000011
01101000100
01101000101
01101000110
01101000111
01101001000
01101001001
01101001010
01101001011
01101001100
01101001101
01101001110
01101001111

Figure 1.17. The 5-tiling shown is generated by 16 different binary sequences of length 11, all beginning with 0110100.

can each be 0 or 1. See Figure 1.17. In general, for $0 \le k \le n - 2$, each k-tiling will be listed 2^{n-2-k} times. In particular, the empty tiling will be listed 2^{n-2} times.

The next identity is based on the fact that for any $t \ge 0$ a tiling can be broken into segments so that all but the last segment have length t or $t + 1$.

Identity 11 *For $m, p, t \ge 0$, $f_{m+(t+1)p} = \sum_{i=0}^{p} \binom{p}{i} f_t^i f_{t-1}^{p-i} f_{m+i}$.*

Question: How many $(m + (t + 1)p)$-tilings exist?

Answer 1: $f_{m+(t+1)p}$.

Answer 2: For any tiling of length $m + (t+1)p$, we break it into $p+1$ segments of length $j_1, j_2, \ldots, j_{p+1}$. For $1 \le i \le p$, $j_i = t$ unless that would result in breaking a domino in half—in which case we let $j_i = t + 1$. Segment $p + 1$ consists of the remaining tiles. Count the number of tilings for which i of the first p segments have length t and the other $p - i$ segments have length $t + 1$. These p segments have total length $it + (p - i)(t + 1) = (t + 1)p - i$. Hence $j_{p+1} = m + i$. Since segments of length t can be covered f_t ways and segments of length $t + 1$ must end with a domino and can be covered f_{t-1} ways, there are exactly $\binom{p}{i} f_t^i f_{t-1}^{p-i} f_{m+i}$ such tilings. See Figure 1.18.

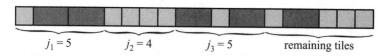

$j_1 = 5 \qquad j_2 = 4 \qquad j_3 = 5 \qquad$ remaining tiles

Figure 1.18. When $t = 4$ and $p = 3$, the tiling above is broken into segments of length $j_1 = 5$, $j_2 = 4$, $j_3 = 5$, and $j_4 = 6$.

The next identity reads better when stated in terms of the traditional definition of Fibonacci numbers (where $F_0 = 0$ and $F_1 = 1$ and thus $f_{n-1} = F_n$ for all $n \ge 0$).

Theorem 1 *For $m \ge 1, n \ge 0$, if $m|n$, then $F_m|F_n$.*

Our combinatorial approach allows us to prove more.

Theorem 2 *For $m \geq 1, n \geq 0$, if m divides n, then f_{m-1} divides f_{n-1}. In fact, if $n = qm$, then $f_{n-1} = f_{m-1} \sum_{j=1}^{q} f_{m-2}^{j-1} f_{n-jm}$.*

Question: When $n = qm$, how many $(n-1)$-tilings exist?

Answer 1: f_{n-1}.

Answer 2: Condition on the smallest j for which the tiling is breakable at cell $jm - 1$. Such a j exists and has value at most q since the tiling is breakable at cell $n-1 = qm-1$. Given j, there are $j-1$ dominoes ending at cells $m, 2m, \ldots, (j-1)m$. The cells preceding these dominoes can be tiled in f_{m-2}^{j-1} ways. Cells $(j-1)m + 1, (j-1)m + 2, \ldots, (jm-1)$ can be tiled f_{m-1} ways. The rest of the board can then be tiled f_{n-jm} ways. See Figure 1.19.

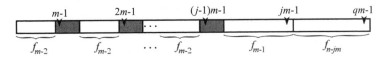

Figure 1.19. There are $f_{m-2}^{j-1} f_{m-1} f_{n-jm}$ ways to tile an $(n-1)$-board when j is the smallest integer for which the tiling is breakable at $jm - 1$.

1.3   A Fun Application

Although the application in this section is not proved entirely by combinatorial means, it utilizes some of the identities from this chapter. Since we have done most of the work to prove it already, it would be a shame to omit it.

For integers a and b, the greatest common divisor, denoted by $\gcd(a, b)$, is the largest positive number dividing both a and b. It is easy to see that for any integer x,

$$\gcd(a, b) = \gcd(b, a - bx), \tag{1.1}$$

since any number that divides both a and b must also divide b and $a - bx$, and vice versa. Two special cases are frequently invoked:

$$\gcd(a, b) = \gcd(b, a - b) \tag{1.2}$$

and

Theorem 3 (Euclidean Algorithm) *If $n = qm + r$, then $\gcd(n, m) = \gcd(m, r)$.*

In the Euclidean algorithm we typically choose $q = \lfloor \frac{n}{m} \rfloor$, so that $0 \leq r < m$. For example, when we apply the Euclidean algorithm to find $\gcd(255, 68)$, we get

$$\gcd(255, 68) = \gcd(68, 51) = \gcd(51, 17) = \gcd(17, 0) = 17.$$

It immediately follows that consecutive Fibonacci numbers are relatively prime, that is

Lemma 4 *For $n \geq 1$, $\gcd(F_n, F_{n-1}) = 1$.*

Proof. This is the world's fastest proof by induction. When $n = 1$, $\gcd(F_1, F_0) = \gcd(1, 0) = 1$. Assuming the lemma holds for the number n, then using (1.2), we get

$$\gcd(F_{n+1}, F_n) = \gcd(F_n, F_{n+1} - F_n) = \gcd(F_n, F_{n-1}) = 1. \qquad \diamond$$

Next we exploit Identity 3 to obtain

Lemma 5 *For $m, n \geq 0$, $F_{m+n} = F_{m+1}F_n + F_m F_{n-1}$.*

Proof. $F_{m+n} = f_{m+(n-1)} = f_m f_{n-1} + f_{m-1} f_{n-2} = F_{m+1}F_n + F_m F_{n-1}.$ $\qquad \diamond$

Finally, we recall that Theorem 1 states if m divides n, then F_m divides F_n. We are now ready to state and prove one of the most beautiful properties of Fibonacci numbers.

Theorem 6 *For $m \geq 1$, $n \geq 0$, $\gcd(F_n, F_m) = F_{\gcd(n,m)}$.*

Proof. Suppose $n = qm + r$, where $0 \leq r < m$. By Lemma 5, $F_n = F_{qm+r} = F_{qm+1}F_r + F_{qm}F_{r-1}$. Thus

$$\gcd(F_n, F_m) = \gcd(F_m, F_{qm+1}F_r + F_{qm}F_{r-1})$$

but by (1.1), we can subtract multiples of F_m from the second term and not change the greatest common divisor. Since by Theorem 1, F_{qm} is a multiple of F_m, it follows that

$$\gcd(F_n, F_m) = \gcd(F_m, F_{qm+1}F_r) = \gcd(F_m, F_r), \qquad (1.3)$$

where the last equality follows since F_m (a divisor of F_{qm}) is relatively prime to F_{qm+1} by Lemma 4.

But what do we have here? Equation (1.3) is the same as the Euclidean Algorithm, but with F's on top. Thus, for example,

$$\gcd(F_{255}, F_{68}) = \gcd(F_{68}, F_{51}) = \gcd(F_{51}, F_{17}) = \gcd(F_{17}, F_0) = F_{17},$$

since $F_0 = 0$. The theorem immediately follows. $\qquad \diamond$

For the reader interested in seeing even more advanced Fibonacci identities, we recommend reading Chapters 2 and 9. One of the treats in store is a proof of Binet's Formula, an exact formula for the nth Fibonacci number. Specifically

$$F_n = \frac{1}{\sqrt{5}}\left[\left(\frac{1+\sqrt{5}}{2}\right)^n - \left(\frac{1-\sqrt{5}}{2}\right)^n\right].$$

An eager reader actually has all the tools necessary to tackle the combinatorial proof and can jump straight to Identity 240.

1.4   Notes

Fibonacci numbers have a long and rich history. They have served as mathematical inspiration and amusement since Leonardo Pisano (filius de Bonacci) first posed his original rabbit reproduction question at the beginning of the 13th century. Fibonacci numbers have touched the lives of mathematicians, artists, naturalists, musicians and more. For a peek at their history, we recommend Ron Knott's impressive web site, Fibonacci Numbers and the Golden Section [32]. Extensive collections of Fibonacci identities are available in Vajda's *Fibonacci & Lucas Numbers, and the Golden Section: Theory and Applications* [58] and Koshy's *Fibonacci and Lucas Numbers with Applications* [33].

The Fibonacci Society is a professional organization focusing on Fibonacci numbers and related mathematics, emphasizing new results, research proposals, challenging problems, and new proofs of old ideas. They publish a professional journal, *The Fibonacci Quarterly,* and organize a biennial international conference.

Combinatorial interpretations of Fibonacci numbers have existed for a long time and can be surveyed in Basin and Hoggatt's article [1] in the inaugural issue of the *Fibonacci Quarterly* or Stanley's *Enumerative Combinatorics Vol. 1* Chapter 1 exercise 14 [51]. We've chosen the tiling interpretation and notation presented in Brigham et. al. [15] and further developed in [8].

Finally, a bijective proof of Cassini's formula similar to the one given for Identity 8 without tilings was given by Werman and Zeilberger [60].

1.5 Exercises

Prove each of the identities below by a direct combinatorial argument.

Identity 12 *For $n \geq 1$, $f_1 + f_3 + \cdots + f_{2n-1} = f_{2n} - 1$.*

Identity 13 *For $n \geq 0$, $f_n^2 + f_{n+1}^2 = f_{2n+2}$.*

Identity 14 *For $n \geq 1$, $f_n^2 - f_{n-2}^2 = f_{2n-1}$.*

Identity 15 *For $n \geq 0$, $f_{2n+2} = f_{n+1}f_{n+2} - f_{n-1}f_n$.*

Identity 16 *For $n \geq 2$, $2f_n = f_{n+1} + f_{n-2}$.*

Identity 17 *For $n \geq 2$, $3f_n = f_{n+2} + f_{n-2}$.*

Identity 18 *For $n \geq 2$, $4f_n = f_{n+2} + f_n + f_{n-2}$.*

Identity 19 demonstrates how any four consecutive Fibonacci numbers generate a Pythagorean Triple.

Identity 19 *For $n \geq 1$,*

$$(f_{n-1}f_{n+2})^2 + (2f_nf_{n+1})^2 = (f_{n+1}f_{n+2} - f_{n-1}f_n)^2 = (f_{2n+2})^2.$$

Identity 20 *For $n \geq p$, $f_{n+p} = \sum_{i=0}^{p} \binom{p}{i} f_{n-i}$.*

Identity 21 *For $n \geq 0$, $\sum_{k=0}^{n}(-1)^k f_k = 1 + (-1)^n f_{n-1}$.*

Identity 22 *For $n \geq 0$, $\displaystyle\prod_{k=1}^{n}\left(1 + \frac{(-1)^{k+1}}{f_k^2}\right) = \frac{f_{n+1}}{f_n}$.*

Identity 23 *For $n \geq 0$, $f_0 + f_3 + f_6 + \cdots + f_{3n} = \frac{1}{2}f_{3n+2}$.*

Identity 24 *For $n \geq 1$, $f_1 + f_4 + f_7 + \cdots + f_{3n-2} = \frac{1}{2}(f_{3n} - 1)$.*

Identity 25 *For $n \geq 1$, $f_2 + f_5 + f_8 + \cdots + f_{3n-1} = \frac{1}{2}(f_{3n+1} - 1)$.*

Identity 26 *For $n \geq 0$, $f_0 + f_4 + f_8 + \cdots + f_{4n} = f_{2n}f_{2n+1}$.*

Identity 27 *For $n \geq 1$, $f_1 + f_5 + f_9 + \cdots + f_{4n-3} = f_{2n-1}^2$.*

Identity 28 *For $n \geq 1$, $f_2 + f_6 + f_{10} + \cdots + f_{4n-2} = f_{2n-1}f_{2n}$.*

Identity 29 *For $n \geq 1$, $f_3 + f_7 + f_{11} + \cdots + f_{4n-1} = f_{2n-1}f_{2n+1}$.*

Identity 30 *For $n \geq 0$, $f_{n+3}^2 + f_n^2 = 2f_{n+1}^2 + 2f_{n+2}^2$.*

Identity 31 *For $n \geq 1$, $f_n^4 = f_{n+2}f_{n+1}f_{n-1}f_{n-2} + 1$.*

There are many combinatorial interpretations for Fibonacci numbers. Show that the interpretations below are equivalent to tiling a board with squares and dominoes by creating a one-to-one correspondence.

1. For $n \geq 0$, f_{n+1} counts binary n-tuples with no consecutive 0s.

2. For $n \geq 0$, f_{n+1} counts subsets S of $\{1, 2, \ldots, n\}$ such that S contains no two consecutive integers.

3. For $n \geq 2$, f_{n-2} counts tilings of an n-board where all tiles have length 2 or greater.

4. For $n \geq 1$, f_{n-1} counts tilings of an n-board where all tiles have odd length.

5. For $n \geq 1$, f_n counts the ways to arrange the numbers 1 through n so that for each $1 \leq i \leq n$, the ith number is $i - 1$ or i or $i + 1$.

6. For $n \geq 0$, f_{2n+1} counts length n sequences of 0s, 1s, and 2s where 0 is never followed immediately by 2.

7. For $n \geq 1$, $f_{2n-1} = \sum a_1 a_2 \cdots a_r$, where $r \geq 1$ and $a_1, \ldots, a_r$ are positive integers that sum to n. For example, $f_5 = 3 + 2 \cdot 1 + 1 \cdot 2 + 1 \cdot 1 \cdot 1 = 8$. (Hint: $a_1 a_2 \cdots a_r$ counts n-tilings with tiles of any length, where a_j is the length of the jth tile, and one cell covered by each tile is highlighted.)

8. For $n \geq 1$, f_{2n} counts $\sum 2^{\text{number of } a_i \text{ that equal } 1}$, summed over the same set as before. For example, when $n = 3 = 2 + 1 = 1 + 2 = 1 + 1 + 1$, $f_6 = 2^0 + 2^1 + 2^1 + 2^3 = 13$.

9. For $n \geq 1$, f_{n+1} counts binary sequences $(b_1, b_2, \ldots, b_n)$, where $b_1 \leq b_2 \geq b_3 \leq b_4 \geq b_5 \cdots$.

Uncounted Identities

The identities listed below are in need of combinatorial proof.

1. For $n \geq 1$, $f_0^3 + f_1^3 + \cdots + f_n^3 = \dfrac{f_{3n+4} + (-1)^n 6 f_{n-1} + 5}{10}$.

2. For $n \geq 0$, $f_1 + 2f_2 + \cdots + nf_n = (n+1)f_{n+2} - f_{n+4} + 3$.

3. There are identities for mf_n analogous to Identities 16–18 for every integer m.

   (a) For $n \geq 4$, $5f_n = f_{n+3} + f_{n-1} + f_{n-4}$.

   (b) For $n \geq 4$, $6f_n = f_{n+3} + f_{n+1} + f_{n-4}$.

   (c) For $n \geq 4$, $7f_n = f_{n+4} + f_{n-4}$.

   (d) For $n \geq 4$, $8f_n = f_{n+4} + f_n + f_{n-4}$.

   (e) For $n \geq 4$, $9f_n = f_{n+4} + f_{n+1} + f_{n-2} + f_{n-4}$.

   (f) For $n \geq 4$, $10f_n = f_{n+4} + f_{n+2} + f_{n-2} + f_{n-4}$.

   (g) For $n \geq 4$, $11f_n = f_{n+4} + f_{n+2} + f_n + f_{n-2} + f_{n-4}$.

   (h) For $n \geq 6$, $12f_n = f_{n+5} + f_{n-1} + f_{n-3} + f_{n-6}$.

   These identities are examples of Zeckendorf's Theorem which states that every integer can be uniquely written as the sum of nonconsecutive Fibonacci numbers. The coefficients in the above formulas are the same as in the unique expansion of positive integers in nonconsecutive integer powers of $\phi = (1+\sqrt{5})/2$. For example $5 = \phi^3 + \phi^{-1} + \phi^{-4}$ and $6 = \phi^3 + \phi^1 + \phi^{-4}$. Is there a unifying combinatorial approach for all of these identities?

4. For $n \geq 4$, $f_n^3 + 3f_{n-3}^3 + f_{n-4}^3 = 3f_{n-1}^2 + 6f_{n-2}^3$. Jay Cordes has shown us a combinatorial proof that requires breaking the tiling triples into over a dozen different cases. Does something simpler exist?

5. Find a combinatorial interpretation for the *Fibonomial coefficient*

$$\binom{n}{m}_F = \frac{(n!)_F}{(m!)_F((n-m)!)_F},$$

where $(0!)_F = 1$, and for $k \geq 1$, $(k!)_F = F_k F_{k-1} \cdots F_1$.

CHAPTER 2

Gibonacci and Lucas Identities

Definition The *Gibonacci numbers* G_n are defined by nonnegative integers G_0, G_1 and for $n \geq 2$, $G_n = G_{n-1} + G_{n-2}$.

Definition The *Lucas numbers* L_n are defined by $L_0 = 2$, $L_1 = 1$ and for $n \geq 2$, $L_n = L_{n-1} + L_{n-2}$.

The first few numbers in the sequence of Lucas numbers are 2, 1, 3, 4, 7, 11, 18, 29, 47, 76, 123, 199, ...

In this chapter, we pursue identities involving *Gibonacci numbers*, which is shorthand for generalized Fibonacci numbers. There are many ways to generalize the Fibonacci numbers, and we shall pursue many of these generalizations in the next chapter, but for our purposes, we say a sequence of nonnegative integers $G_0, G_1, G_2, \ldots$ is a Gibonacci sequence if for all $n \geq 2$,

$$G_n = G_{n-1} + G_{n-2}.$$

Of all the Gibonacci sequences, the initial conditions that lead to the most beautiful identities correspond to the Fibonacci and Lucas numbers.

2.1 Combinatorial Interpretation of Lucas Numbers

As we shall see Lucas numbers operate like Fibonacci numbers running in circles. Define ℓ_n to be the number of ways to tile a circular board composed of n labeled cells with curved squares and dominoes. For example $\ell_4 = 7$ as illustrated in Figure 2.1. Clearly there are more ways to tile a *circular n-board* than a straight n-board since it is now possible for a single domino to cover cells n and 1. We define an *n-bracelet* to be a tiling of a circular n-board. A bracelet is *out-of-phase* when a single domino covers cells n and 1 and *in-phase* otherwise. In Figure 2.1, we see that there are five in-phase 4-bracelets and two out-of-phase 4-bracelets. Figure 2.2 illustrates that $\ell_1 = 1$, $\ell_2 = 3$, and $\ell_3 = 4$. Notice that there are two ways to create a 2-bracelet with a single domino—either in-phase or out-of-phase.

From our initial data, the number of n-bracelets looks like the Lucas sequence. To prove that they continue to grow like the Lucas sequence, we must argue that for $n \geq 3$,

$$\ell_n = \ell_{n-1} + \ell_{n-2}.$$

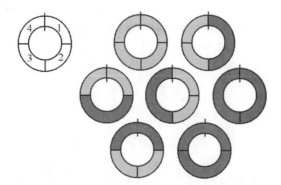

Figure 2.1. A circular 4-board and its seven bracelets. The first five bracelets are in-phase and the last two are out-of-phase.

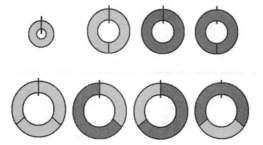

Figure 2.2. There is one 1-bracelet and there are three 2-bracelets and four 3-bracelets.

To see this we simply condition on the *last tile* of the bracelet. We define the *first* tile to be the tile that covers cell 1, which could either be a square, a domino covering cells 1 and 2, or a domino covering cells n and 1. The second tile is the next tile in the clockwise direction, and so on. The last tile is the one that precedes the first tile. Since it is the first tile, not the last, that determines the phase of the tiling, there are ℓ_{n-1} n-bracelets that end with a square and ℓ_{n-2} n-bracelets that end with a domino. By removing the last tile and closing up the resulting gap, we produce smaller bracelets.

To make the recurrence valid for $n = 2$, we define $\ell_0 = 2$, and interpret this to mean that there are two empty tilings of the circular 0-board, an in-phase 0-bracelet and an out-of-phase 0-bracelet. This leads to a combinatorial interpretation of Lucas numbers.

Combinatorial Theorem 2 *For $n \geq 0$, let ℓ_n count the ways to tile a circular n-board with squares and dominoes. Then ℓ_n is the nth Lucas number; that is*

$$\ell_n = L_n.$$

As one might expect, there are many identities with Lucas numbers that resemble Fibonacci identities. In addition, there are many beautiful identities where Lucas and Fibonacci numbers interact.

2.2 Lucas Identities

Identity 32 *For $n \geq 1$, $L_n = f_n + f_{n-2}$.*

Question: How many tilings of a circular n-board exist?

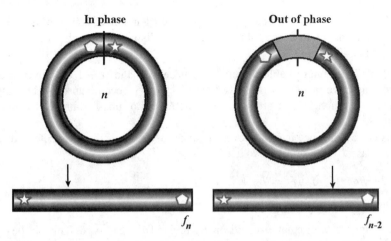

Figure 2.3. Every circular n-bracelet can be reduced to an n-tiling or an $(n-2)$-tiling, depending on its phase.

Answer 1: By Combinatorial Theorem 2, there are L_n n-bracelets.

Answer 2: Condition on whether the tiling is in-phase or out-of-phase. Since an in-phase tiling can be straightened into an n-tiling, there are f_n in-phase bracelets. Likewise, an out-of-phase n-bracelet must have a single domino covering cells n and 1. Cells 2 through $n-1$ can then be covered as a straight $(n-2)$-tiling in f_{n-2} ways. Hence the total number of n-bracelets is $f_n + f_{n-2}$. See Figure 2.3.

The next identity associates an odd-length board tiling with a board and bracelet pair.

Identity 33 *For $n \geq 0$, $f_{2n-1} = L_n f_{n-1}$.*

Set 1: Tilings of a $(2n-1)$-board. This set has size f_{2n-1}.

Set 2: Bracelet-tiling pairs (B, T) where the bracelet has length n and the tiling has length $n-1$. This set has size $L_n f_{n-1}$.

Correspondence: Given a $(2n-1)$-board T^*, there are two cases to consider, as illustrated in Figure 2.4.

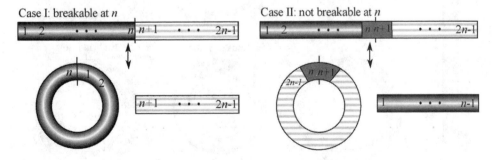

Figure 2.4. A $(2n-1)$-tiling can be converted to an n-bracelet and $(n-1)$-tiling. In our correspondence, the n-bracelet is in-phase if and only if the $(2n-1)$-tiling is breakable at cell n.

Case 1. If T^* is breakable at cell n, then glue the right side of cell n to the left side of cell 1 to create an in-phase n-bracelet B, and cells $n + 1$ through $2n - 1$ form an $(n - 1)$-tiling T.

Case 2. If T^* is unbreakable at cell n, then cells n and $n + 1$ are covered by a domino which we denote by d. Cells 1 through $n - 1$ become an $(n - 1)$-tiling T and cells n through $2n - 1$ are used to create an out-of-phase n-bracelet with d as its first tile.

This correspondence is easily reversed since the phase of the n-bracelet indicates whether Case 1 or Case 2 is invoked.

Identity 34 *For $n \geq 0$, $5f_n = L_n + L_{n+2}$.*

Set 1: Tilings of an n-board. This set has size f_n.

Set 2: Tilings of a circular n-board or a circular $(n + 2)$-board. This set has size $L_n + L_{n+2}$.

Correspondence: To prove the identity, we establish a 1-*to*-5 *correspondence* between Set 1 and Set 2. That is, for every tiling in Set 1, we can create five bracelets in Set 2 in such a way that every bracelet in Set 2 is created exactly once. Hence Set 2 is five times as large as Set 1.

Given an n-tiling, four of the five bracelets arise naturally. See Figure 2.5. We can create

1. an in-phase n-bracelet by gluing cell n to cell 1, or
2. an in-phase $(n + 2)$-bracelet ending with two inserted squares, or
3. an in-phase $(n + 2)$-bracelet ending with an inserted domino, or
4. an out-of-phase $(n + 2)$-bracelet ending with an inserted domino.

At this point we pause to investigate which bracelets have not yet been created. We are missing out-of-phase n-bracelets and $(n + 2)$-bracelets that end with a square

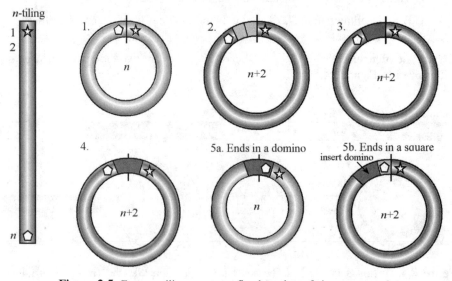

Figure 2.5. Every n-tiling generates five bracelets of size n or $n + 2$.

preceded by a domino. So the fifth bracelet depends on whether the original n-tiling ends with a square or domino. If it ends with a domino, we create

5a. an out-of-phase n-bracelet by rotating bracelet 1 clockwise by one cell.

If it ends with a square, we create

5b. an in-phase $(n+2)$-bracelet that ends with the square preceded by an inserted domino.

See Figure 2.5.

Identity 35 *For $n \geq 0$,*

$$\sum_{r=0}^{n} f_r L_{n-r} = (n+2) f_n.$$

Set 1: The set of n-tilings. This set has size f_n.

Set 2: The set of ordered pairs (A, B) where A is an r-tiling and B is an $(n-r)$-bracelet for some $0 \leq r \leq n$. This set has size $\sum_{r=0}^{n} f_r L_{n-r}$.

Correspondence: We provide a 1-to-$(n+2)$ correspondence between Set 1 and Set 2. Given an n-tiling X, we first examine for each $1 \leq r \leq n-1$, whether or not X is breakable at cell r. See Figure 2.6. If so, then $X = AB$ where A is an r-tiling and B is an $(n-r)$-tiling, and we associate the tiling pair (A, B) where B is an in-phase $(n-r)$-bracelet. Otherwise, $X = AdB$ where A is an $(r-1)$-tiling, B is an

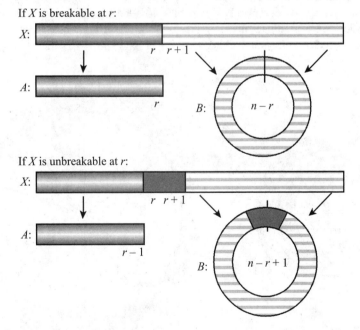

Figure 2.6. Given an n-tiling, for every cell $0 \leq r \leq n-1$ we generate a tiling-bracelet pair (A, B) whose lengths sum to n. The nth cell generates two tiling pairs, since there are two empty 0-bracelets.

$(n - r - 1)$-tiling. We associate the tiling pair (A, dB) where dB is an out-of-phase $(n - r + 1)$-bracelet. This accounts for $n - 1$ tiling pairs. We also associate (with $r = 0$) the tiling pair $(\emptyset, X)$, where X is an in-phase n-bracelet and (with $r = n$) the two tiling pairs $(X, \emptyset^+)$ and $(X, \emptyset^-)$ since there are two 0-bracelets, one in-phase and one out-of phase. Altogether, each n-tiling generates $(n + 2)$ tiling pairs (A, B). The process is easily reversed by examining the phase of bracelet B.

Identity 36 *For* $n \geq 0$, $L_n^2 = L_{2n} + (-1)^n \cdot 2$.

Set 1: The set of ordered pairs of concentric n-bracelets. This set has size L_n^2.

Set 2: The set of $2n$-bracelets.

Correspondence: We present an almost one-to-one correspondence between Set 1 and Set 2 for the case where n is odd. We leave the even case for the reader.

Since n is odd, each n-bracelet must contain at least 1 square, and therefore the concentric bracelets must contain a *first fault* at some cell $1 \leq k \leq n$. That is, both bracelets are breakable at cell k, but not at cells $1, 2, \ldots, k - 1$. From this we create a $2n$-bracelet as follows. Starting with the tile covering cell 1 of the outer bracelet (so that the new bracelet has the same phase as the outer bracelet) we tile cells 1 through k as in the outer bracelet, then tile cells $k + 1$ through $k + n$ using the tiles that cover cells $k + 1, k + 2, \ldots, n, 1, 2, \ldots, k$ of the inner bracelet. Finally, we tile cells $k + n + 1$ through $2n$ using the remaining tiles of the outer bracelet. See Figure 2.7. The resulting $2n$-bracelet has the property that a *diameter* can pass through it, entering between cells k and $k + 1$ and departing between cells $n + k$ and $n + k + 1$. Further no diameter exists that enters between j and $j + 1$ for $1 \leq j \leq k - 1$. Hence this process is completely reversible provided that a diameter exists in our $2n$-bracelet. Since a square at cell j in our $2n$-bracelet guarantees that a diameter can enter between cell j and either cell $j - 1$ or cell $j + 1$, there are precisely

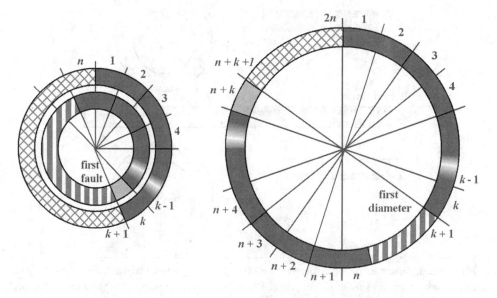

Figure 2.7. Concentric n-bracelets producing a $2n$-bracelet.

two tilings of the $2n$-bracelet that have no diameters, namely the all-domino tilings. (Note that the all domino tilings have no diameter since n is odd.) Consequently, $L_n^2 = L_{2n} - 2$.

2.3 Combinatorial Interpretation of Gibonacci Numbers

To see how to interpret Gibonacci numbers combinatorially, we take a second look at Lucas numbers. From Combinatorial Theorem 2, we know that L_n counts the number of ways to tile an n-bracelet with squares and dominoes. Notice that we can "straighten out" an n-bracelet, by writing it as an n-tiling starting with the first tile (the tile covering cell 1) with one caveat. The caveat is that if the first tile is a domino, we need to indicate whether it is an in-phase or out-of-phase domino. For example, the seven 4-bracelets of Figure 2.1 have been straightened out in *phased tilings* in Figure 2.8. Summarizing, L_n counts the number of *phased n-tilings* where an initial domino has **two** possible phases and an initial square has **one** possible phase. The next theorem should then come as no surprise.

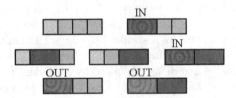

Figure 2.8. The seven 4-bracelets can be straightened out to become "phased" 4-bracelets.

Combinatorial Theorem 3 *Let $G_0, G_1, G_2, \ldots$ be a Gibonacci sequence with nonnegative integer terms. For $n \geq 1$, G_n counts the number of n-tilings, where the initial tile is assigned a phase. There are G_0 choices for a domino phase and G_1 choices for a square phase.*

Proof. Let a_n denote the number of phased n-tilings with G_0 and G_1 phases for initial dominoes and squares, respectively. Clearly, $a_1 = G_1$. A phased 2-tiling consists of either a phased domino (G_0 choices) or a phased square followed by an unphased square (G_1 choices). Hence $a_2 = G_0 + G_1 = G_2$. To see that a_n grows like Gibonacci numbers we merely condition on the last tile, which immediately gives us $a_n = a_{n-1} + a_{n-2}$.   ◇.

In order for our theorem to be valid when $n = 0$, we combinatorially define the number of phased 0-tilings to be G_0, the number of domino phases.

2.4 Gibonacci Identities

Elementary Gibonacci Identities

Using this combinatorial interpretation of G_n, many identities become transparent. For instance, by conditioning on the first tile of a phased tiling (see Figure 2.9), it immediately follows that

Identity 37 *For $n \geq 1$, $G_n = G_0 f_{n-2} + G_1 f_{n-1}$.*

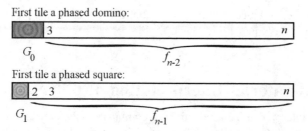

Figure 2.9. A phased n-tiling either begins with a phased domino or a phased square.

The next identity is a generalization of Identity 3 from Chapter 1.

Identity 38 *For $m \geq 1$, $n \geq 0$, $G_{m+n} = G_m f_n + G_{m-1} f_{n-1}$.*

Question: How many phased $(m + n)$-tilings exist?

Answer 1: By definition, there are G_{m+n} such tilings.

Answer 2: Condition on whether the phased $(m + n)$-tiling is breakable at cell m. See Figure 2.10. The number of breakable tilings is $G_m f_n$ since such a tiling consists of a phased m-tiling followed by a standard n-tiling. The number of unbreakable tilings is $G_{m-1} f_{n-1}$ since such tilings contain a phased $(m-1)$-tiling, followed by a domino covering cells m and $m+1$, followed by a standard $(n-1)$-tiling. Note if $m = 1$, then the phase of the 0-tiling is applied to the domino covering cells 1 and 2. Altogether, there are $G_m f_n + G_{m-1} f_{n-1}$ $(m + n)$-tilings.

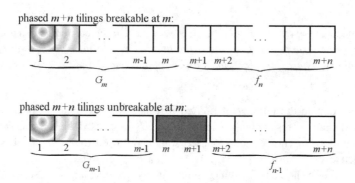

Figure 2.10. A phased $(m + n)$-tiling is either breakable or unbreakable at cell m.

Identity 39 *For $n \geq 0$, $\sum_{k=0}^{n} G_k = G_{n+2} - G_1$.*

Question: How many phased $(n + 2)$-tilings contain at least one domino?

Answer 1: There are G_{n+2} phased $(n+2)$-tilings including the tilings consisting of only squares. So there are $G_{n+2} - G_1$ tilings with at least one domino.

Answer 2: Condition on the location of the last domino. For $0 \leq k \leq n$, there are G_k tilings where the last domino covers cells $k+1$ and $k+2$ as illustrated in Figure 2.11. Notice that when the last domino covers cells 1 and 2, it must have one of G_0 phases. So the argument is still valid.

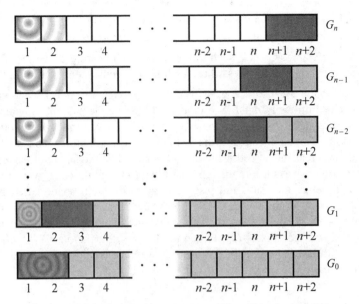

Figure 2.11. Conditioning on the location of the last domino.

The next identity is a generalization of Identity 6 of Chapter 1.

Identity 40 *For* $n \geq p \geq 0$, $G_{n+p} = \sum_{i=0}^{p} \binom{p}{i} G_{n-i}$.

Question: How many phased $(n+p)$-tilings exist?

Answer 1: G_{n+p}.

Answer 2: Condition on the number of dominoes that appear among the last p tiles. When the last p tiles consist of i dominoes and $p - i$ squares, there are $\binom{p}{i}$ ways to arrange these tiles. The length of these p tiles is $p + i$. The remaining board has length $(n + p) - (p + i) = n - i$ and can be tiled G_{n-i} ways. See Figure 2.12.

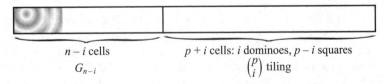

$n - i$ cells $p + i$ cells: i dominoes, $p - i$ squares
G_{n-i} $\binom{p}{i}$ tiling

Figure 2.12. A phased $(n + p)$-tiling with i dominoes among the last p tiles.

Similar identities are presented in the exercises.

Simultaneous Tilings

The identities in this subsection are a little trickier since they involve tiling two boards simultaneously. It will be convenient to think of one board as the "top board" and the other as the "bottom board". We begin with

Identity 41 *For* $n \geq 0$, $\sum_{i=1}^{2n} G_i G_{i-1} = G_{2n}^2 - G_0^2$.

Question: In how many ways can two boards of length $2n$ be given phased tilings so that the tiling pair contains at least one square somewhere?

Answer 1: There are $G_{2n}^2 - G_0^2$ such tilings, since each board can be tiled G_{2n} ways and we throw away the G_0^2 cases where both boards consist only of dominoes.

Answer 2: Let the top board consist of cells 1 through $2n$ and the bottom board consist of cells 2 through $2n + 1$. See Figure 2.13. Since the phased tiling pair has at least one square somewhere, there must be at least one fault that goes through both tilings. Condition on the location of the last fault. The last fault occurs at cell i when both tilings are breakable at cell i, but at no future cell. If the last fault occurs at cell $i \geq 2$, there are G_i ways to tile the top board before the fault, G_{i-1} ways to tile the bottom board before the fault, and just one way to tile both boards after the fault. All tiles to the right of the fault are dominoes except for a single square in cell $i + 1$ of the board with an odd tail length. The argument is slightly different when $i = 1$; here the number of pairs is $G_1 G_0$. Hence for $1 \leq i \leq 2n$, there are $G_{i-1} G_i$ phased tiling pairs with last fault at cell i. Altogether, we have $\sum_{i=1}^{2n} G_i G_{i-1}$ phased tiling pairs with at least one square.

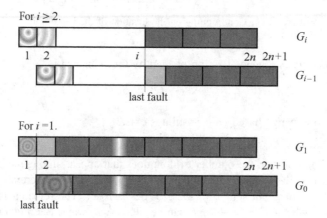

Figure 2.13. A last fault at cell i.

When the board has odd length, a similar argument yields Identity 64 given in the exercises.

The identities of this section were all based on simultaneous tilings of two phased boards with the same initial conditions (determined by G_0 and G_1) for both the top and bottom boards. However, all of these identities can be easily generalized to situations where the initial conditions are not necessarily the same. In what follows, we shall assume that $G_0, G_1, G_2, \ldots$ and $H_0, H_1, H_2, \ldots$ are both Gibonacci sequences, possibly with different initial conditions. For instance, the next identity completely generalizes the sum of the squares of Fibonacci numbers given in Identity 9.

Identity 42 *For $n \geq 1$, $G_0 H_1 + \sum_{i=1}^{2n-1} G_i H_i = G_{2n} H_{2n-1}$.*

Question: In how many ways can a phased $2n$-board and a phased $(2n-1)$-board be tiled, where the first board has G_0 initial domino phases and G_1 initial square phases, while the second board has H_0 initial domino phases and H_1 initial square phases?

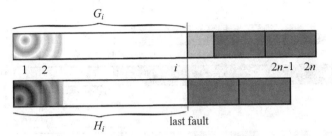

Figure 2.14. There are $G_i H_i$ tilings with last fault at cell i.

Answer 1: $G_{2n}H_{2n-1}$.

Answer 2: We let the top board cover cells 1 through $2n$ and the bottom board cover cells 1 through $2n - 1$. We condition on the last fault, if one exists. Here, since the first board has even length, the only fault-free tilings are those that have all dominoes in both boards, except for a single phased square in the bottom board. Hence there are $G_0 H_1$ fault-free tilings. Otherwise, by the same reasoning as in the last identity, there are $G_i H_i$ tilings whose last fault occurs at cell i, where $1 \leq i \leq 2n - 1$. Altogether our boards may be tiled $G_0 H_1 + \sum_{i=1}^{2n-1} G_i H_i$ ways. See Figure 2.14.

A similar identity results when the longer board has odd length.

Tail swapping identities

The identities of this subsection utilize the tail swapping technique from Chapter 1.

Identity 43 *Let $G_0, G_1, G_2, \ldots$ and $H_0, H_1, H_2, \ldots$ be Gibonacci sequences. Then for $0 \leq m \leq n$, $G_m H_n - G_n H_m = (-1)^m (G_0 H_{n-m} - G_{n-m} H_0)$.*

Set 1: The set of tiling pairs of a phased m-board and a phased n-board, where the initial conditions of the m-board are determined by G_0, G_1, and the initial conditions of the n-board are determined by H_0, H_1. This set has size $G_m H_n$.

Set 2: The set of tiling pairs of a phased n-board and a phased m-board, where the initial conditions of the n-board are determined by G_0, G_1, and the initial conditions of the m-board are determined by H_0, H_1. This set has size $G_n H_m$.

Correspondence: The identity makes sense once we draw the appropriate picture. Place an m-board on top of an n-board, as in Figure 2.15. Tail-swapping provides a one-to-one correspondence between the faulty tilings of Set 1 and Set 2. The number of fault-free tilings depends on the parity of m. As illustrated in Figure 2.15, when m is even, the fault free tilings of Set 1 consist of $m/2$ dominoes in the top board and the bottom board begins with a phased square followed by $m/2$ dominoes, followed by an (unphased) $(n-m-1)$-tiling. Consequently, the number of fault-free tilings of Set 1 is $G_0 H_1 f_{n-m-1}$. By the same logic, there are $G_1 H_0 f_{n-m-1}$ fault-free tilings of Set 2. Consequently, when m is even, the difference in size between Sets 1 and 2 is the number of fault-free tilings of Set 1 minus the number of fault-free tilings of Set 2. That is,

$$G_m H_n - G_n H_m = G_0 H_1 f_{n-m-1} - H_0 G_1 f_{n-m-1}.$$

However, it is combinatorially clear that $H_1 f_{n-m-1} = H_{n-m} - H_0 f_{n-m-2}$ since

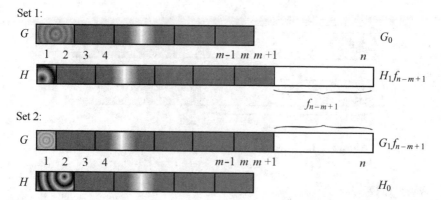

Figure 2.15. When m is even, there are $G_0 H_1 f_{n-m-1}$ fault-free tilings in Set 1, and $G_1 H_0 f_{n-m-1}$ fault-free tilings in Set 2.

both count $(n-m)$-tilings that begin with a phased square. Similarly, $G_1 f_{n-m-1} = G_{n-m} - G_0 f_{n-m-2}$. Thus

$$G_m H_n - G_n H_m = G_0 H_{n-m} - H_0 G_{n-m},$$

since the $G_0 H_0 f_{n-m-2}$ terms conveniently disappear.

At first glance, the next identity looks like a generalization of Identity 43 but it is nothing more than a translation with a change of variables. If we translate the Gibonacci sequence $H_0, H_1, H_2, \ldots$ to H_k, H_{k+1}, H_{k+2}, we obtain another Gibonacci sequence. Then if we substitute $n = m + h$, we obtain

Identity 44 *Let $G_0, G_1, G_2, \ldots$ and $H_0, H_1, H_2, \ldots$ be Gibonacci sequences. Then for $m, h, k \geq 0$, $G_m H_{m+h+k} - G_{m+h} H_{m+k} = (-1)^m (G_0 H_{h+k} - G_h H_k)$.*

The following generalization of Identity 33, will result in a magical application.

Identity 45 *For $0 \leq m \leq n$, $G_{n+m} + (-1)^m G_{n-m} = G_n L_m$.*

Set 1: The set of phased $(n+m)$-tilings. This set has size G_{n+m}.

Set 2: The set of ordered pairs (A, B), where A is a phased n-tiling, and B is an m-bracelet. This set has size $G_n L_m$.

Correspondence: The identity is clearly true when $m = 0$, so we assume $m \geq 1$. We create an almost one-to-one correspondence between these two sets. Let P be a phased $(n+m)$-tiling. If P is breakable at cell n, we create a phased n-tiling A from the phased tiling of the first n cells of P. Using cells $n+1$ through $n+m$ create B, an in-phase m-bracelet, as in Figure 2.16. If P is not breakable at cell n, then create the tiling pair of Figure 2.17, where the top tiling is the phased $(n-1)$-tiling from cells 1 through $n-1$ of P. The bottom tiling is an unphased $(m+1)$-tiling, beginning with a domino, from cells n through $n+m$ of P. Now perform a tail swap, if possible, to create a pair of tilings with sizes n and m, where the n-tiling is phased, and the m-tiling is unphased, but begins with a domino. These become a phased tiling and out-of-phase bracelet in the natural way.

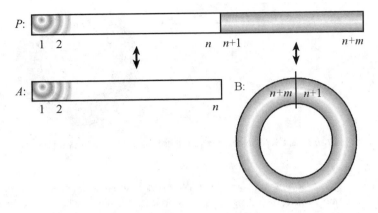

Figure 2.16. Breakable phased $(n+m)$-tilings naturally become phased n-tilings with an in-phase m-bracelet.

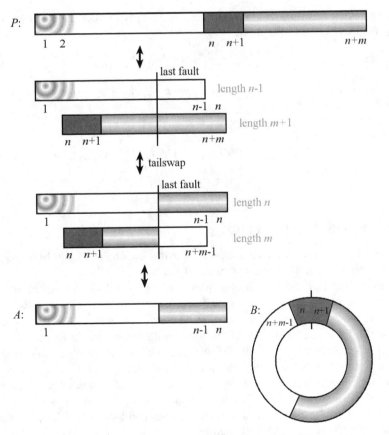

Figure 2.17. Unbreakable phased $(n + m)$-tilings become phased n-tilings with an out-of-phase m-bracelet.

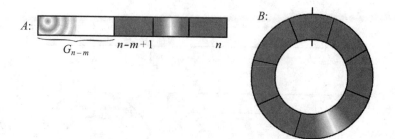

Figure 2.18. When m is even, these pairs are unachievable.

When is tail swapping not possible? When m is even, the $(m + 1)$-tiling must have at least one square, resulting in at least one fault. Thus when m is even, we can always tail swap, but there are G_{n-m} unachievable tiling pairs where the bottom m-tiling consists of all dominoes and the phased n-tiling has only dominoes in cells $n-m+1$ through n. See Figure 2.18. Thus when m is even, $G_nL_m = G_{n+m}+G_{n-m}$ as desired. By a similar argument, when m is odd, $G_{n+m} = G_nL_m + G_{n-m}$.

The following identities are consequences of Identity 44, but can be proved directly as well. We leave these to the reader.

Identity 46 *For* $n \geq 1$, $G_{n+1}G_{n-1} - G_n^2 = (-1)^n(G_1^2 - G_0G_2)$.

Identity 47 *For* $0 \leq m \leq n$, $H_{n-m} = (-1)^m(F_{m+1}H_n - F_mH_{n+1})$.

Identity 48 *For* $n \geq 1$ *and* $0 \leq m \leq n$,

$$G_{n+m} - (-1)^mG_{n-m} = F_m(G_{n-1} + G_{n+1}).$$

A Gibonacci Magic Trick

Let's take a break from combinatorial proofs for just a moment. If you have made it this far into the book, we reward you with a bit of *mathemagics*.

The mathemagician hands a sheet of paper as in Figure 2.19 to a volunteer and says, "Secretly write a positive integer in Row 1 and another positive integer in Row 2. Next, add those numbers together and put the sum in Row 3. Add Row 2 to Row 3 and place the answer in Row 4. Continue in this fashion until numbers are in Rows 1 through 10. Now using a calculator, if you wish, add all the numbers in Rows 1 through 10 together." While the spectator is adding, the mathemagician glances at the sheet of paper for just a second, then instantly reveals the total. "Now using a calculator, divide the number in Row 10 by the number in Row 9, and announce the first three digits of your answer. What's that you say? 1.61? Now turn over the paper and look what I have written." The back of the paper says "I predict the number 1.61".

The explanation of this trick involves nothing more than high school algebra. For the first part, observe in Figure 2.20 that if Row 1 contains x and Row 2 contains y then the total of Rows 1 through 10 will sum to $55x + 88y$. As luck would have it, (actually

1	
2	
3	
4	
5	
6	
7	
8	
9	
10	
TOTAL	

Figure 2.19. Enter a Gibonacci sequence, with positive integers in Rows 1 and 2.

1	x
2	y
3	$x + y$
4	$x + 2y$
5	$2x + 3y$
6	$3x + 5y$
7	**5x + 8y**
8	$8x + 13y$
9	$13x + 21y$
10	$21x + 34y$
TOTAL	**55x + 88y**

Figure 2.20. The sum of the 10 numbers is Row 7 times 11.

by the next identity), the number in Row 7 is $5x + 8y$. Consequently, the grand total is simply 11 times Row 7.

As for the ratio, it's all about adding fractions badly. For any two fractions $\frac{a}{b} < \frac{c}{d}$ with positive numerators and denominators, the quantity $\frac{a+c}{b+d}$ is called the *mediant* (sometimes called the *freshman sum*) and it's easy to show that

$$\frac{a}{b} < \frac{a + c}{b + d} < \frac{c}{d}.$$

Consequently, the ratio of (Row 10)/(Row 9) satisfies

$$1.615\ldots = \frac{21}{13} = \frac{21x}{13x} < \frac{21x + 34y}{13x + 21y} < \frac{34y}{21y} = \frac{34}{21} = 1.619\ldots.$$

The first part of this trick was a special case of the following identity, which is an immediate consequence of Identities 39 and 45.

Identity 49 *For $n \geq 0$, $\sum_{i=0}^{4n+1} G_i = G_{2n+2}L_{2n+1}$.*

2.5 Notes

Edouard Lucas (pronounced LOO-KAH) was the first person to call the sequence $0, 1, 1, 2,$ $3, 5, 8, \ldots$, the Fibonacci sequence. Combinatorial interpretations of Lucas numbers appear in [8, 20, 44, 54] including independent sets of vertices in cycle graphs, circular binary sequences with no consecutive zeros, and tilings that are not allowed to begin and end with a domino.

A combinatorial interpretation for Generalized Fibonacci numbers appears in [13]. Other generalizations of Fibonacci numbers will appear in the next chapter.

2.6 Exercises

Prove each of the identities below by a direct combinatorial argument.

Identity 50 *For $n \geq 2$, $L_n = f_{n-1} + 2f_{n-2}$.*

Identity 51 *For $n \geq 0$, $f_{n-1} + L_n = 2f_n$.*

Identity 52 *For $n \geq 0$, $5f_n = L_{n+1} + 2L_n$.*

Identity 53 *For $n \geq 0$, $5f_n^2 = L_{n+1}^2 + 4(-1)^n$.*

Identity 54 *For $n \geq 1$, $L_1^2 + L_3^2 + \cdots + L_{2n-1}^2 = f_{4n-1} - 2n$.*

Identity 55 *For $n \geq 0$, $L_{2n+1} - L_{2n-1} + L_{2n-3} - L_{2n-5} + \cdots \pm L_3 \mp L_1 = f_{2n+1}$.*

Identity 56 *For $n \geq 2$, $L_n^4 = L_{n-2}L_{n-1}L_{n+1}L_{n+2} + 25$.*

Identity 57 *For $n \geq 0$, $\sum_{r=0}^{n} L_r L_{n-r} = (n+1)L_n + 2f_n$.*

Identity 58 *For $n \geq 2$, $5 \sum_{r=0}^{n-2} f_r f_{n-2-r} = nL_n - f_{n-1}$.*

Identity 59 *For $n \geq 2$, $G_n^4 = G_{n+2}G_{n+1}G_{n-1}G_{n-2} + (G_2 G_0 - G_1^2)^2$.*

Identity 60 *For $n \geq 1$, $L_n^2 = L_{n+1}L_{n-1} + (-1)^n \cdot 5$.*

Identity 61 *For $n \geq 0$, $\sum_{k=1}^{n} G_{2k-1} = G_{2n} - G_0$.*

Identity 62 *For $n \geq 0$, $G_1 + \sum_{k=1}^{n} G_{2k} = G_{2n+1}$.*

Identity 63 *For $m, p, t \geq 0$, $G_{m+(t+1)p} = \sum_{i=0}^{p} \binom{p}{i} f_t^i f_{t-1}^{p-i} G_{m+i}$.*

Identity 64 *For $i \geq 1$, $\sum_{i=1}^{n-1} G_{i-1}G_{i+2} = G_n^2 - G_1^2$.*

Identity 65 *For $n \geq 2$, $G_{n+1}^2 = 4G_{n-1}G_n + G_{n-2}^2$.*

Identity 66 *For $n \geq 1$, $\sum_{i=1}^{n-1} G_{i-1}G_{i+2} = G_n^2 - G_1^2$.*

Identity 67 *For $n \geq 1$, $G_0 G_1 + \sum_{i=1}^{n-1} G_i^2 = G_n G_{n-1}$.*

Identity 68 *Let $G_0, G_1, G_2, \ldots$ and $H_0, H_1, H_2, \ldots$ be Gibonacci sequences, then for $1 \leq m \leq n$, $G_m H_n - G_{m-1}H_{n+1} = (-1)^m [G_0 H_{n-m+2} - G_1 H_{n-m+1}]$.*

Identity 69 *$G_{n+1} + G_n + G_{n-1} + 2G_{n-2} + 4G_{n-3} + 8G_{n-4} + \cdots + 2^{n-1}G_0 = 2^n(G_0 + G_1)$.*

Identity 70 *For $n \geq 0$, $G_{n+3}^2 + G_n^2 = 2G_{n+1}^2 + 2G_{n+2}^2$.*

Other Exercises

1. Translating a Gibonacci sequence $G_0, G_1, G_2, \ldots$ by m gives another Gibonacci sequence $G_m, G_{m+1}, G_{m+2}, \ldots$. Show how to derive Identity 38 from Identity 37 by translating G_i to G_{i+m}.

2. There are many combinatorial interpretations for Lucas numbers. Show that the interpretations below are equivalent to tiling a circular board with squares and dominoes by creating a one-to-one correspondence.

    (a) For $n \geq 2$, L_n counts circular binary sequences with no consecutive 0s.

    (b) For $n \geq 1$, L_n counts tilings of an $(n + 1)$-board that do not begin and end with a domino.

CHAPTER **3**

Linear Recurrences

Definition Given integers $c_1, \ldots, c_k$, a *kth order linear recurrence* is defined by $a_0, a_1, \ldots,$ a_{k-1}, and for $n \geq k$, $a_n = c_1 a_{n-1} + c_2 a_{n-2} + \cdots + c_k a_{n-k}$.

Definition Given integers s and t, the *Lucas sequence of the first kind* is defined by $U_0 = 0$, $U_1 = 1$ and for $n \geq 2$, $U_n = sU_{n-1} + tU_{n-2}$. For combinatorial convenience, we also define for $n \geq -1$, $u_n = U_{n+1}$. When $s = t = 1$, these are the Fibonacci numbers: $U_n = F_n$, and $u_n = f_n$.

Definition Given integers s and t, the *Lucas sequence of the second kind* is defined by $V_0 = 2$, $V_1 = s$ and for $n \geq 2$, $V_n = sV_{n-1} + tV_{n-2}$. When $s = t = 1$, these are the Lucas numbers: $V_n = L_n$.

Generalizing the Fibonacci Numbers

The recurrence for Fibonacci numbers can be extended in many different directions. The phrase "Generalized Fibonacci Number" has been used to describe numbers generated by such recurrences as

$$a_n = a_{n-1} + a_{n-2}$$
$$a_n = a_{n-1} + a_{n-2} + \cdots + a_{n-k}$$
$$a_n = a_{n-1} + a_{n-k}$$
$$a_n = sa_{n-1} + ta_{n-2}$$

with various assumptions about the initial conditions. In this chapter, we shall go even further and combinatorially explain numbers generated by kth order linear recurrences

$$a_n = c_1 a_{n-1} + c_2 a_{n-2} + \cdots + c_k a_{n-k}$$

where $c_1, c_2, \ldots, c_k$ are nonnegative integers. We shall first describe these under the "ideal" initial conditions for combinatorial purposes, then we deal with other initial conditions. By the end of the chapter, we will even have combinatorial interpretations for identities where the coefficients $c_1, \ldots, c_k$ are negative, irrational or complex numbers, with *any* initial conditions.

3.1   Combinatorial Interpretations of Linear Recurrences

When the initial conditions are just right, kth order linear recurrences have an especially simple combinatorial interpretation.

Combinatorial Theorem 4 *Let $c_1, c_2, \ldots, c_k$ be nonnegative integers. Let $u_0, u_1, \ldots$ be the sequence of numbers defined by the following recurrence: For $n \geq 1$,*

$$u_n = c_1 u_{n-1} + c_2 u_{n-2} + \cdots + c_k u_{n-k} \qquad (3.1)$$

with "ideal" initial conditions $u_0 = 1$, and for $j < 0$, $u_j = 0$. Then for all $n \geq 0$, u_n counts **colored tilings** *of an n-board, with tiles of length at most k, where for $1 \leq i \leq k$, each tile of length i is assigned one of c_i colors.*

Proof. The theorem is clearly true for $n \leq 0$. We proceed by the now familiar argument of conditioning on the last tile. For $n \geq 1$ and $1 \leq i \leq k$, an n-board having a final colored tile of length i occurs $c_i u_{n-i}$ ways. Summing over all possible choices of i yields the recurrence in (3.1).   ◇

When $k = 2$, the ideal initial conditions generate Lucas numbers of the first kind and Combinatorial Theorem 4 reduces to

Combinatorial Theorem 5 *Let s, t be nonnegative integers. Suppose $u_0 = 1$, $u_1 = s$ and for $n \geq 2$,*

$$u_n = s u_{n-1} + t u_{n-2}. \qquad (3.2)$$

Then for all $n \geq 0$, u_n counts colored tilings of an n-board with squares and dominoes, where there are s colors for squares and t colors for dominoes.

When $s = t = 1$, u_n is a Fibonacci number: $u_n = f_n = F_{n-1}$.

In the last chapter, we saw how closely the Lucas numbers interacted with the Fibonacci numbers. In the same way, there exists a natural companion sequence for u_n, the Lucas numbers of the second kind. When $s = t = 1$, V_n is the traditional Lucas number: $V_n = L_n$. We leave it to the reader to prove the following.

Combinatorial Theorem 6 *Let s, t be nonnegative integers. Suppose $V_0 = 2$, $V_1 = s$ and for $n \geq 2$,*

$$V_n = s V_{n-1} + t V_{n-2}. \qquad (3.3)$$

Then for all $n \geq 0$, V_n counts **colored bracelets** *of length n, where there are s colors for squares and t colors for dominoes.*

Nearly all of the identities of the first two chapters generalize seamlessly by just adding a splash of color. For example, the Fibonacci–Lucas identity $f_{2n-1} = f_{n-1} L_n$ is unchanged in the colorized version.

Identity 71 *For $n \geq 1$, $u_{2n-1} = u_{n-1} V_n$.*

The proof is exactly the same as in Identity 33. We simply work with colored squares and dominoes instead of uncolored ones.

Recall Identity 34 from Chapter 2: For $n \geq 0$, $5f_n = L_n + L_{n+2}$. In our proof, every n-tiling generated five bracelets, and every bracelet of length n or length $n+2$ was created exactly once. In the colorized version,

Identity 72 *For $n \geq 0$, $(s^2 + 4t)u_n = tV_n + V_{n+2}$.*

The proof is structured exactly as in Identity 34, but now each colored n-tiling generates $s^2 + 4t$ bracelets. In the process, every colored n-bracelet is generated t times and every colored $(n + 2)$-bracelet is generated exactly once. For more details, see the appendix of solutions. Incidentally, the quantity $s^2 + 4t$ shows up frequently in generalized Fibonacci and Lucas identities, and is called the *discriminant*.

In Chapter 8, we prove that even the greatest common divisor property for Fibonacci numbers generalizes without a hitch: Let s, t be relatively prime nonnegative integers and let U_m, U_n be Lucas numbers of the first kind.

$$\gcd(m, n) = g \implies \gcd(U_m, U_n) = U_g.$$

For more identities, see the exercises at the end of this chapter.

How do we accommodate arbitrary initial conditions? As with the Gibonacci numbers of the last chapter, the initial conditions introduce a *phase* for the initial tile. We begin with second-order linear recurrences.

Combinatorial Theorem 7 *Let s, t, a_0, a_1 be nonnegative integers, and for $n \geq 2$, define*

$$a_n = sa_{n-1} + ta_{n-2}.$$

*For $n \geq 1$, a_n counts the ways to tile an n-board with squares and dominoes where each tile, **except the initial one** has a color. There are s colors for squares and t colors for dominoes. The initial tile is given a **phase**; there are a_1 phases for an initial square and ta_0 phases for an initial domino.*

Proof. The number of ways to tile a 1-board (using a single phased square) is a_1. The number of ways to tile a 2-board (with two squares or a single phased domino) is $sa_1 + ta_0 = a_2$. Proceeding inductively, and conditioning on the last tile, we see that, for $n \geq 2$, the number of ways to tile an n-board is $sa_{n-1} + ta_{n-2} = a_n$, as desired. ◇

We call such a tiling a *phased colored tiling*. For consistency, we say that a 0-board has a_0 phased colored tilings.

Finally, we generalize Combinatorial Theorem 4 to accommodate other initial conditions.

Combinatorial Theorem 8 *Let $c_1, c_2, \ldots, c_k, a_0, a_1, \ldots, a_{k-1}$ be nonnegative integers, and for $n \geq k$, define*

$$a_n = c_1 a_{n-1} + c_2 a_{n-2} + \cdots + c_k a_{n-k}. \tag{3.4}$$

If the initial conditions satisfy

$$a_i \geq \sum_{j=1}^{i-1} c_j a_{i-j} \tag{3.5}$$

for $1 \leq i \leq k$, then for $n \geq 1$, a_n counts the ways to tile an n-board using colored tiles of length at most k, where each tile, except the initial one, has a color. Specifically, for $1 \leq i \leq k$, each tile of length i may be assigned any of c_i different colors, but an initial tile of length i is assigned one of p_i phases, where

$$p_i = a_i - \sum_{j=1}^{i-1} c_j a_{i-j}. \tag{3.6}$$

Proof. By inequality (3.5), p_i is a nonnegative integer for $1 \leq i \leq k$. For $1 \leq n \leq k$, a phased colored tiling of an n-board consists of either a single phased tile of length n, or it ends with a colored tile of length i for some $1 \leq i \leq n-1$. Thus the number of such tilings is equal to

$$p_n + \sum_{i=1}^{n-1} c_i a_{n-i} = \left(a_n - \sum_{j=1}^{n-1} c_j a_{n-j} \right) + \sum_{i=1}^{n-1} c_i a_{n-i} = a_n.$$

For $n > k$, there must be more than one tile. Hence the last tile (say of length i) can be assigned one of c_i colors, and is preceded by a phased colored tiling of length $n-i$. Hence the number of phased colored n-tilings satisfies recurrence (3.4), as desired.   ◇

Notice for a kth order linear recurrence equations (3.6) and (3.4) imply that

$$p_k = a_k - \sum_{i=1}^{k-1} c_i a_{k-i} = c_k a_0, \tag{3.7}$$

which is always nonnegative. Thus inequality (3.5) is always valid for $i = k$ and obviously also true for $i = 1$. This is why no restrictions are imposed on the initial conditions of second-order recurrences.

Although for $k > 2$, Combinatorial Theorem 8 requires that the initial conditions satisfy inequality (3.5), the identities we prove depending on inequality (3.5) will be valid for all initial conditions. The proof of this can be derived using linear algebra as presented in [4], or by the combinatorial explanation at the end of this chapter.

Notice that the ideal initial conditions of Combinatorial Theorem 4 were chosen so that for each $1 \leq i \leq k$, $p_i = u_i - \sum_{j=1}^{i-1} c_j u_{i-j} = c_i u_0 = c_i$. Hence an initial tile of length i has as many choices as any other tile of length i. Thus u_n counts the number of *un*phased colored tilings of length n, as asserted by Combinatorial Theorem 4.

3.2 Identities for Second-Order Recurrences

The identities and proofs in this section are generalizations of identities presented in Chapter 2. Here we consider sequences $a_0, a_1, a_2, \ldots$ generated by the recurrence: for $n \geq 2$, $a_n = s a_{n-1} + t a_{n-2}$, where a_0 and a_1 are arbitrary nonnegative integers. By Combinatorial Theorem 7, a_n counts the number of phased colored n-tilings, where an initial square has $p_1 = a_1$ possible phases, and an initial domino has $p_2 = t a_0$ possible phases. We let u_n count the number of unphased colored n-tilings, where u_n is generated by the same recurrence, where $u_0 = 1$ and $u_1 = s$. We assume that s and t are nonnegative integers. Notice that when $s = t = 1$, $a_n = G_n$, the nth Gibonacci number and $u_n = f_n$.

Our first identity generalizes Identity 38, $G_{m+n} = G_m f_n + G_{m-1} f_{n-1}$.

Identity 73 *For $m, n \geq 1$, $a_{m+n} = a_m u_n + t a_{m-1} u_{n-1}$.*

Question: How many phased colored $(m+n)$-tilings exist (where, as usual, we have a_1 phases for an initial square and $t a_0$ phases for an initial domino)?

Answer 1: By Combinatorial Theorem 7, there are a_{m+n} such tilings.

phased colored $m+n$ tilings breakable at m:

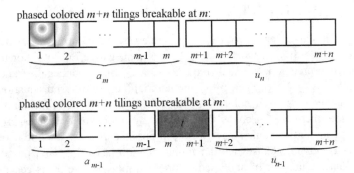

phased colored $m+n$ tilings unbreakable at m:

Figure 3.1. A phased colored $(m + n)$-tiling is either breakable or unbreakable at cell m.

Answer 2: Condition on whether the phased colored $(m + n)$-tiling is breakable at cell m. See Figure 3.1. The number of breakable tilings is $a_m u_n$ since such a tiling consists of a phased colored m-tiling followed by an unphased colored n-tiling. The number of unbreakable tilings is $a_{m-1} t u_{n-1}$ since such tilings contain a phased $(m-1)$-tiling, followed by a colored domino covering cells m and $m+1$, followed by an unphased colored $(n - 1)$-tiling. (The argument when $m = 1$ is a little different.) Altogether, there are $a_m u_n + t a_{m-1} u_{n-1}$ phased colored $(m + n)$-tilings.

Next we generalize Identity 39 for (unphased) colored tilings where $a_0 = 1, a_1 = s$, and for $n \geq 2$, $a_n = s a_{n-1} + t a_{n-2}$.

Identity 74 *For* $n \geq 2$,
$$a_n - 1 = (s - 1)a_{n-1} + (s + t - 1)[a_0 + a_1 + \cdots + a_{n-2}].$$

Question: How many colored n-tilings exist, excluding the tiling consisting of all white squares?

Answer 1: Since $a_0 = 1$, $a_1 = s$ is equivalent to the ideal initial conditions, there are $a_n - 1$ such tilings.

Answer 2: Here we partition our tilings according to the last tile that is not a white square. Suppose the last tile that is not a white square begins on cell k. If $k = n$, the last tile is a square and there are $s - 1$ choices for its color. There are a_{n-1} colored tilings that can precede it for a total of $(s - 1)a_{n-1}$ tilings ending in a nonwhite square. If $1 \leq k \leq n - 1$, the tile covering cell k can be any colored domino or a nonwhite square. There are $s + t - 1$ ways to pick this tile and the previous cells can be tiled a_{k-1} ways. Altogether, there are $(s - 1)a_{n-1} + \sum_{k=1}^{n-1}(s + t - 1)a_{k-1}$ colored n-tilings, as desired.

Notice how easily the argument generalizes if we partition according to the last tile that is not a square of color 1 or 2 or ... or c. Then the same reasoning gives us

Identity 75 *For any* $1 \leq c \leq s$, *for* $n \geq 0$,
$$a_n - c^n = (s - c)a_{n-1} + ((s - c)c + t)[a_0 c^{n-2} + a_1 c^{n-3} + \cdots + a_{n-2}].$$

As with noncolored tilings, identities that count pairs of colored tilings can still be obtained by conditioning on the last fault and tail swapping. See the exercises for more examples.

3.3   Identities for Third-Order Recurrences

Here we consider sequences $a_0, a_1, a_2, \ldots$ generated by the recurrence: for $n \geq 3$, $a_n = c_1 a_{n-1} + c_2 a_{n-2} + c_3 a_{n-3}$, where a_0, a_1, a_2 are nonnegative integers. For the simplest combinatorial interpretation, we shall require c_1, c_2, c_3 to be nonnegative and to satisfy inequality (3.5), $a_2 \geq c_1 a_1$. Then by Combinatorial Theorem 8 and equation (3.7), a_n counts the number of phased colored n-tilings with squares, dominoes and (length 3) trominoes where there are c_1 colors for squares, c_2 colors for dominoes, c_3 colors for trominoes, and an initial square has $p_1 = a_1$ possible phases, an initial domino has $p_2 = a_2 - c_1 a_1$ possible phases and an initial tromino has $p_3 = c_3 a_0$ possible phases. We let u_n count the number of unphased colored n-tilings, where u_n is generated by the same recurrence and $u_0 = 1$, $u_1 = c_1$, and $u_2 = c_1^2 + c_2$.

Identity 76 *For $m, n \geq 2$, $a_{m+n} = a_m u_n + c_2 a_{m-1} u_{n-1} + c_3 (a_{m-2} u_{n-1} + a_{m-1} u_{n-2})$.*

Question: How many phased colored $(m + n)$-tilings exist using squares, dominoes and trominoes where there are c_1 colors for squares, c_2 colors for dominoes, c_3 colors for trominoes, and an initial square has $p_1 = a_1$ possible phases, an initial domino has $p_2 = a_2 - c_1 a_1$ possible phases and an initial tromino has $p_3 = c_3 a_0$ possible phases?

Answer 1: By Combinatorial Theorem 8, there are a_{m+n} such tilings.

Answer 2: Condition on the length of the tile (if any) covering cells m and $m+1$. See Figure 3.2. As in the proof of Identity 73, the number of breakable tilings is $a_m u_n$, and the number of ways for cells m and $m+1$ to be covered by the same domino is $a_{m-1} c_2 u_{n-1}$. There are two ways that cells m and $m+1$ can be covered by the same tromino. As illustrated in Figure 3.2, this can occur $a_{m-2} c_3 u_{n-1} + a_{m-1} c_3 u_{n-2}$

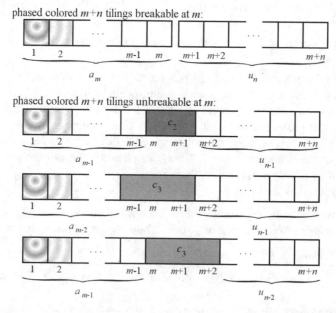

Figure 3.2. A phased colored $(m + n)$-tiling is either breakable at cell m, has a domino covering cells m and $m + 1$, or has one of two types of trominoes covering cells m and $m + 1$.

ways. Altogether, there are $a_m u_n + c_2 a_{m-1} u_{n-1} + c_3(a_{m-2} u_{n-1} + a_{m-1} u_{n-2})$ phased colored $(m+n)$-tilings.

The next identity is a generalization of an identity satisfied by the **Tribonacci numbers** (defined by $T_n = T_{n-1} + T_{n-2} + T_{n-3}$, where $T_0 = T_1 = 1$, $T_2 = 2$):

$$\sum_{i=1}^{n} T_i = \frac{1}{2}(T_{n+2} + T_n - 3).$$

More generally, for third-order recurrences, we have:

Identity 77 *For $n \geq 0$,*

$$c_1^n(c_1 a_2 + c_3 a_0) + (c_1 c_2 + c_3) \sum_{i=1}^{n} c_1^{n-i} a_i = c_1 a_{n+2} + c_3 a_n.$$

Question: How many phased, colored $(n+3)$-tilings exist using c_1 colored squares, c_2 colored dominoes and c_3 colored trominoes with initial phases as described in Combinatorial Theorem 8 and that end with a square or a tromino?

Answer 1: By conditioning on the last tile, there are $c_1 a_{n+2} + c_3 a_n$ such tilings.

Answer 2: Condition on the location of the last domino or tromino. For $1 \leq i \leq n$, we count phased colored $(n+3)$-tilings whose last domino or tromino begins at cell $i+1$. Notice that our ending condition disallows the possibility of a last domino beginning at cell $n+2$. There are $(c_1 c_2 + c_3) c_1^{n-i} a_i$ such tilings since cells $i+1, i+2, i+3$ consists of either a single tromino or a single domino followed by a square ($c_3 + c_2 c_1$ choices), the tiles following cell $i+3$ must be squares (c_1^{n-i} choices) and cells 1 through i may be tiled arbitrarily (a_i choices). The only uncounted tilings are those that begin with a phased tile, followed by all squares. There are $p_3 c_1^n = c_3 a_0 c_1^n$ such tilings that begin with a tromino. Otherwise, we have a 2-tiling followed by all squares, which can be done $a_2 c_1^{n+1}$ ways. Hence these tilings can be created in $c_1^n(c_3 a_0 + c_1 a_2)$ ways. Altogether, our tilings can be constructed $c_1^n(c_1 a_2 + c_3 a_0) + (c_1 c_2 + c_3) \sum_{i=1}^{n} c_1^{n-i} a_i$ ways.

We end this section with special types of integer sequences determined by third-order linear recurrences. The **3-bonacci numbers** are defined by $u_n = u_{n-1} + u_{n-3}$, with

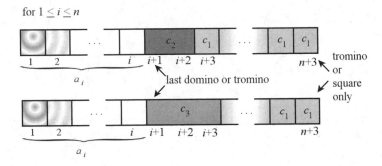

Figure 3.3. Conditioning on the last domino or tromino, when the last tile must be a square or tromino.

ideal initial conditions $u_0 = 1$ and $u_j = 0$ for $j < 0$. The first few 3-bonacci numbers are $1, 1, 1, 2, 3, 4, 6, 9, 13, 19, 28, 41, 60, 88, 129, 189, \ldots$. The next identity extends the argument of Identity 4 in Chapter 1.

Identity 78 *Let u_n be defined for $n \geq 1$ by $u_n = u_{n-1} + u_{n-3}$ where $u_0 = 1$ and $u_j = 0$ for $j < 0$. Then for $n \geq 0$,*

$$\sum_{i \geq 0} \binom{n - 2i}{i} = u_n.$$

Question: How many unphased, uncolored n-tilings exist that use only squares and trominoes?

Answer 1: By Combinatorial Theorem 4, u_n counts the number of unphased uncolored tilings of an n-board with squares and trominoes.

Answer 2: Condition on the number of trominoes. Provided that $i \leq n/3$, an n-board with i trominoes must have $n - 3i$ squares, and therefore $n - 2i$ tiles altogether. There are $\binom{n-2i}{i}$ ways to arrange these tiles.

For a colorful generalization, we consider *generalized* 3-bonacci numbers.

Identity 79 *Let u_n be defined for $n \geq 1$ by $u_n = su_{n-1} + tu_{n-3}$ where $u_0 = 1$ and $u_j = 0$ for $j < 0$. Then*

$$\sum_{i \geq 0} \binom{n - 2i}{i} t^i s^{n-3i} = u_n.$$

Question: How many unphased, colored n-tilings exist where we have s colors for squares and t colors for trominoes?

Answer 1: By Combinatorial Theorem 4, u_n counts the number of unphased colored tilings of an n-board with squares and trominoes.

Answer 2: Condition on the number of trominoes. Provided that $i \leq n/3$, an n-board with i trominoes must have $n - 3i$ squares, and therefore $n - 2i$ tiles altogether. There are $\binom{n-2i}{i}$ ways to arrange these tiles, then $t^i s^{n-3i}$ ways to assign them colors.

We conclude this section with one more 3-bonacci identity, extending Identity 5 from Chapter 1.

Identity 80 *Let u_n be defined for $n \geq 1$ by $u_n = u_{n-1} + u_{n-3}$ where $u_0 = 1$ and $u_j = 0$ for $j < 0$. Then*

$$\sum_{a=0}^{n} \sum_{b=0}^{n} \sum_{c=0}^{n} \binom{n - b - c}{a} \binom{n - a - c}{b} \binom{n - a - b}{c} = u_{3n+2}.$$

Question: How many unphased, uncolored $(3n + 2)$-tilings exist using only squares and trominoes?

Answer 1: u_{3n+2}.

Answer 2: The number of squares in any tiling of a $(3n+2)$-board must be 2 greater than a multiple of 3. Hence there will exist two *goalpost* squares, say located at cells

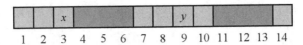

$$1 \quad 2 \quad 3 \quad 4 \quad 5 \quad 6 \quad 7 \quad 8 \quad 9 \quad 10 \quad 11 \quad 12 \quad 13 \quad 14$$

Figure 3.4. The number of ways to arrange the tiles within the three regions defined by the goalposts x and y is $\binom{2}{0}\binom{3}{1}\binom{3}{1}$.

x and y, such that there are an equal number of squares to the left of x, strictly between x and y, and to the right of y. For example in a tiling with eight squares, x and y are the cells occupied by the third and sixth square. See Figure 3.4. We condition on the number of trominoes in the three regions defined by the goal posts. If the number of trominoes in each region is, from left to right, a, b, c, then there are a total of $a + b + c$ trominoes and $(3n + 2) - 3(a + b + c)$ squares, including the two goalpost squares. Hence each region has $n - (a + b + c)$ squares. The leftmost region has $n - b - c$ tiles, a of which are trominoes, and there are $\binom{n-b-c}{a}$ ways to arrange them. Likewise the tiles of the second and third region can be arranged $\binom{n-a-c}{b}$ ways and $\binom{n-a-b}{c}$ ways, respectively.

As a, b, and c vary, we obtain the total number of $(3n + 2)$-tilings as

$$\sum_{a=0}^{n}\sum_{b=0}^{n}\sum_{c=0}^{n} \binom{n-b-c}{a}\binom{n-a-c}{b}\binom{n-a-b}{c}.$$

3.4 Identities for kth Order Recurrences

While all of the previous identities can be generalized to kth order recurrences, some of which are explored in the exercises, we finish this section with two identities that are particularly elegant.

Identity 81 *Let g_n be the kth order Fibonacci sequence defined by $g_j = 0$ for $j < 0$, $g_0 = 1$, and for $n \geq 1$, $g_n = g_{n-1} + g_{n-2} + \cdots + g_{n-k}$. Then for all integers n,*

$$g_n = \sum_{n_1}\sum_{n_2}\cdots\sum_{n_k} \frac{(n_1 + n_2 + \cdots + n_k)!}{n_1! n_2! \cdots n_k!},$$

where the summation is over all nonnegative integers $n_1, n_2, \ldots, n_k$ such that $n_1 + 2n_2 + \cdots + kn_k = n$.

Question: How many unphased, uncolored n-tilings exist where we are allowed tiles of lengths at most k.

Answer 1: By Combinatorial Theorem 4, g_n.

Answer 2: Condition on the number of tiles of each length. If for $1 \leq i \leq k$ there are n_i tiles of length i, then we must have $n_1 + 2n_2 + \cdots + kn_k = n$. The number of ways to permute these tiles is given by the *multinomial coefficient*

$$\binom{n_1 + n_2 + \cdots + n_k}{n_1, n_2, \ldots, n_k} = \frac{(n_1 + n_2 + \cdots + n_k)!}{n_1! n_2! \cdots n_k!}.$$

Finally, we generalize Identity 80 to k-**bonacci numbers**.

Identity 82 *Let u_n be the k-bonacci number defined for $n \geq 1$ by $u_n = u_{n-1} + u_{n-k}$, where $u_0 = 1$ and for $j < 0$, $u_j = 0$. Then for $n \geq 0$, $u_{kn+(k-1)}$ equals*

$$\sum_{x_1=0}^{n} \sum_{x_2=0}^{n} \cdots \sum_{x_k=0}^{n} \binom{n-(x_2+x_3+\cdots+x_k)}{x_1} \binom{n-(x_1+x_3+\cdots+x_k)}{x_2} \cdots \binom{n-(x_1+x_2+\cdots+x_{k-1})}{x_k}.$$

Question: How many unphased uncolored $(kn + (k-1))$-tilings exist where we are only allowed squares and (length k) k-ominoes?

Answer 1: $u_{kn+(k-1)}$.

Answer 2: The number of squares in any tiling of a $(kn + (k-1))$-board must be $k-1$ greater than a multiple of k. Now there will exist $k-1$ goalpost squares dividing our board into k regions, each containing an equal number of squares. For $1 \leq i \leq k$, let x_i denote the number of k-ominoes in region i, then the number of squares in each region will be $n-(x_1+x_2+\cdots+x_k)$. The number of ways to permute the tiles in region i is given by the binomial coefficient $\binom{n-(x_1+x_2+\cdots+x_k)+x_i}{x_i}$, as desired.

3.5   Get Real! Arbitrary Weights and Initial Conditions

In this chapter, we have developed a combinatorial interpretation for any sequence of numbers a_n defined by a kth order linear recurrence. To summarize, we were given initial conditions $a_0, a_1, \ldots, a_{k-1}$, and for $n \geq k$,

$$a_n = c_1 a_{n-1} + c_2 a_{n-2} + \cdots + c_k a_{n-k}.$$

By Combinatorial Theorem 8, a_n counts the number of phased, colored tilings of length n where each tile, except for the first one, is assigned a color. For $1 \leq i \leq k$, a length i tile can be assigned one of c_i colors. The initial tile is assigned a phase. For $1 \leq i \leq k$, an initial tile of length i is assigned one of p_i phases where $p_i = a_i - \sum_{j=1}^{i-1} c_j a_{i-j}$.

The above interpretation only makes sense when $a_0, \ldots, a_{k-1}, c_1, \ldots, c_k$, and $p_1, \ldots, p_k$ are nonnegative integers. And yet, most of the identities proved in this book remain true when these quantities are negative or irrational or complex numbers (or from any commutative ring). This section illustrates how combinatorial arguments can still be used to overcome these apparent obstacles.

Suppose that every c_i, a_i and p_i is chosen from the set of complex numbers. (Some of them *could* be nonnegative integers, but we don't require it.) Instead of assigning a discrete number of colors for each tile, we assign weights. For $1 \leq i \leq k$, tiles of length i have weight c_i except for the initial tile which has weight p_i as defined above. We define the *weight of an n-tiling* to be the product of the weights of its individual tiles. For example, the 13-tiling "tromino-square-domino-domino-square-tromino-square" has weight $(c_1)^3(c_2)^2(c_3)^2$ with ideal initial conditions and has weight $p_3(c_1)^3(c_2)^2(c_3)$ with arbitrary initial conditions. By essentially the same argument as used in proving Combinatorial Theorem 8, it follows that for $n \geq 1$, a_n is the sum of the weights of all weighted n-tilings, which we call the *total weight* of an n-board.

If X is an m-tiling of weight w_X and Y is an n-tiling of weight w_Y, then X and Y can be glued together to create an $(m+n)$-tiling of weight $w_X w_Y$. If an m-board can be tiled s different ways and has total weight $a_m = w_1 + w_2 + \cdots + w_s$ and an n-board can be tiled t ways with total weight $a_n = x_1 + x_2 + \cdots + x_t$, then the sum of the weights of all weighted $(m + n)$-tilings breakable at cell m is

$$\sum_{i=1}^{s} \sum_{j=1}^{t} w_i x_j = (w_1 + w_2 + \cdots + w_s)(x_1 + x_2 + \cdots + x_t) = a_m a_n.$$

Now we are prepared to revisit some of our previous identities using the weighted approach. For Identity 73 where squares have weight s and dominoes have weight t, we find the total weights of an $(m + n)$-board in two different ways. By definition, the total weight is a_{m+n}. On the other hand, the total weight is comprised of the total weight of those tilings that are breakable at cell m $(a_m u_n)$ plus the total weight of those tilings that are unbreakable at cell m $(a_{m-1} t u_{n-1})$.

For identities like Identity 87, we define the weight of a tiling pair to be the product of the weights of all the tiles, and define the total weight as before. Next we observe that tail swapping preserves the weight of the tiling pair since no tiles are created or destroyed in the process. Consequently, the total weight of the set of *faulty* tiling pairs (X, Y) where X and Y are n-tilings equals the total weight of the faulty tiling pairs (X', Y'), where X' is an $(n + 1)$-tiling and Y' is an $(n - 1)$-tiling. The fault free tiling pair, for the even and odd case, will consist of n dominoes and therefore have weight t^n. Hence Identity 87 remains true even when s and t are *non*-nonnegative integers.

3.6 Notes

Some of this material originally appeared in [4] and [10], and was developed with undergraduate Chris Hanusa. Identities 104 and 105 were originally proved by algebraic means in [59]. Exercises 3 and 4 were suggested to us by Peter G. Anderson. Exercise 5 is an alternate colored tiling interpretation of kth order linear recurrences which avoids the restrictions on a_i given by inequality (3.5) suggested to us by Dan Velleman and Bill Zwicker.

For an analytic approach to the same material, consult [30].

3.7 Exercises

Please provide combinatorial proofs for the identities below.

For Identities 83 through 96, u_n and V_n are defined by equations (3.2) and (3.3). That is, s and t are nonnegative integers, $u_0 = 1$, $u_1 = s$, $V_0 = 2$, $V_1 = s$, and for $n \geq 2$, $u_n = s u_{n-1} + t u_{n-2}$, and $V_n = s V_{n-1} + t V_{n-2}$.

Identity 83 *For $n \geq 2$, $V_n = u_n + t u_{n-2}$.*

Identity 84 *For $n \geq 0$,*

$$u_{2n+1} = s[u_0 + u_2 + \cdots + u_{2n}] + (t - 1)[u_1 + u_3 + \cdots + u_{2n-1}].$$

Identity 85 *For $n \geq 0$,*

$$u_{2n} - 1 = s[u_1 + u_3 + \cdots + u_{2n-1}] + (t - 1)[u_0 + u_2 + \cdots + u_{2n-2}].$$

Identity 86 *For $n \geq 0$,*

$$s \sum_{k=0}^{n} u_k^2 t^{n-k} = u_n u_{n+1}.$$

Identity 87 *For $n \geq 0$, $u_n^2 = u_{n+1}u_{n-1} + (-1)^n t^n$.*

Identity 88 *For $n \geq r \geq 1$, $u_n^2 - u_{n-r}u_{n+r} = (-t)^{n-r+1}u_{r-1}^2$.*

Identity 89 *For $n \geq 1$, $V_n^2 = V_{n+1}V_{n-1} + (s^2 + 4t)(-t)^n$.*

Identity 90 *For $n \geq r \geq 1$, $V_n^2 = V_{n+r}V_{n-r} + (s^2 + 4t)(-t)^{n-r+1}u_{r-1}^2$.*

Identity 91 *For $n \geq 0$, $V_{2n} = V_n^2 - 2(-t)^n$.*

Identity 92 *For $n \geq 1$, $2u_n = su_{n-1} + V_n$.*

Identity 93 *For $n \geq 0$, $(s^2 + 4t)u_n + sV_{n+1} = 2V_{n+2}$.*

Identity 94 *For $m \geq 0$, $n \geq 1$, $2u_{m+n} = u_m V_n + V_{m+1}u_{n-1}$.*

Identity 95 *For $m, n \geq 0$, $2V_{m+n} = V_m V_n + (s^2 + 4t)u_{m-1}u_{n-1}$.*

Identity 96 *For $n \geq 0$, $V_n^2 = (s^2 + 4t)u_{n-1}^2 + 4(-t)^n$.*

For Identities 97 through 103, find combinatorial proofs for the quantities defined by the second-order recurrence $a_n = sa_{n-1} + ta_{n-2}$, for $n \geq 2$, where s, t, a_0 and a_1 are given nonnegative integers.

Identity 97 *For $n \geq 0$, $a_{2n}^2 - t^{2n}a_0^2 = s \sum_{i=1}^{2n} t^{2n-i}a_{i-1}a_i$.*

Identity 98 *For $n \geq 0$, $a_{2n+1}^2 - a_1^2 t^{2n} = a_{2n+2}a_{2n} - a_0^2 t^{2n+1} - a_0 a_1 s t^{2n}$.*

Identity 99 *For $n \geq 0$, $t \sum_{k=0}^{n} s^{n-k}a_k = a_{n+2} - s^{n+1}a_1$.*

Identity 100 *For $n \geq 0$, $a_{2n+1} = a_1 t^n + s \sum_{k=1}^{n} t^{n-k}a_{2k}$.*

Identity 101 *For $n \geq 0$, $a_{2n} = a_0 t^n + s \sum_{k=1}^{n} t^{n-k}a_{2k-1}$.*

Identity 102 *For $n \geq 1$,*

$$a_{2n+1} = s(a_0 + a_2 + \cdots + a_{2n}) + (t-1)(a_1 + a_3 + \cdots + a_{2n-1}).$$

Identity 103 *For $n \geq 1$,*

$$a_{2n} - 1 = s(a_1 + a_3 + \cdots + a_{2n-1}) + (t-1)(a_0 + a_2 + \cdots + a_{2n-2}).$$

For the next two identities, we are given nonnegative integers $c_1, c_2, c_3, a_0, a_1, a_2$ and for $n \geq 3$, the third-order recurrence $a_n = c_1 a_{n-1} + c_2 a_{n-2} + c_3 a_{n-3}$. Find combinatorial proofs of these. (For combinatorial convenience, you may assume that $a_2 - c_1 a_1 \geq 0$.)

Identity 104 *For $n \geq 1$,*

$$c_2^n(a_2 - c_1 a_1) + (c_1 c_2 + c_3) \sum_{i=1}^{n} c_2^{n-i}a_{2i-1} = c_2 a_{2n} + c_3 a_{2n-1}.$$

Identity 105 *For $n \geq 1$,*

$$c_2^{n-1}(c_2 a_1 + c_3 a_0) + (c_1 c_2 + c_3) \sum_{i=1}^{n-1} c_2^{n-1-i}a_{2i} = c_2 a_{2n-1} + c_3 a_{2n-2}.$$

Other Exercises

1. Prove Combinatorial Theorem 6.

2. What recurrence and initial conditions generate colored bracelets, where for $1 \leq i \leq k$, there are c_i color choices for a tile of length i?

3. Let $u_n = 0$ for $j < 0$, and for $n \geq 1$, $u_n = u_{n-2} + u_{n-3}$. Let $w_n = 0$, for $j < 0$, $w_0 = w_1 = w_2 = 1$, $w_3 = w_4 = 2$, and for $n \geq 5$, $w_n = w_{n-1} + w_{n-5}$. Combinatorially prove, for $n \geq 2$, $w_n = u_{n+2}$.

4. Using the following second-order and third-order recurrences,

$$f_n = f_{n-1} + f_{n-2}, \text{ with } f_0 = 1, f_1 = 1,$$

$$g_n = g_{n-1} + g_{n-3}, \text{ with } g_0 = 1, g_1 = 1, g_2 = 1,$$

$$h_n = h_{n-2} + h_{n-3}, \text{ with } h_0 = 1, h_1 = 0, h_2 = 1,$$

$$t_n = t_{n-1} + t_{n-2} + t_{n-3}, \text{ with } t_0 = 1, t_1 = 1, t_2 = 2,$$

prove for $n \geq 0$,

   (a) $t_{n+3} = f_{n+3} + \sum_{p+q=n} f_p t_q$,
   (b) $t_{n+2} = g_{n+2} + \sum_{p+q=n} g_p t_q$,
   (c) $t_{n+1} = h_{n+1} + \sum_{p+q=n} h_p t_q$.

5. For the linear recurrence $a_n = c_1 a_{n-1} + \cdots + c_k a_{n-k}$ with nonnegative initial conditions $a_0, a_1, \ldots, a_k$, prove that a_n counts the number of *restricted* phased colored tilings, where an initial tile of length i is assigned one of a_i phases, and each subsequent tile of length i is assigned one of c_i colors. The restriction is that the first tile must have length ℓ where $0 \leq \ell \leq k - 1$ and the second tile, if there is one, must cover cell k.

CHAPTER 4

Continued Fractions

Definition Given integers $a_0 \geq 0, a_1 \geq 1, a_2 \geq 1, \ldots, a_n \geq 1$, define $[a_0, a_1, \ldots, a_n]$ to be the fraction in lowest terms for

$$a_0 + \cfrac{1}{a_1 + \cfrac{1}{a_2 + \cfrac{1}{\ddots + \cfrac{1}{a_n}}}}.$$

For example, $[2, 3, 4] = \frac{30}{13}$.

4.1 Combinatorial Interpretation of Continued Fractions

You might be surprised to learn that the finite continued fraction

$$3 + \cfrac{1}{7 + \cfrac{1}{15 + \cfrac{1}{1 + \cfrac{1}{292}}}} \qquad \text{and its reversal} \qquad 292 + \cfrac{1}{1 + \cfrac{1}{15 + \cfrac{1}{7 + \cfrac{1}{3}}}}$$

have the same numerator. These fractions simplify to $\frac{103993}{33102}$ and $\frac{103993}{355}$ respectively. In this chapter, we provide a combinatorial interpretation for the numerators and denominators of continued fractions which makes this reversal phenomenon easy to see. Our interpretation also allows us to visualize many important identities involving continued fractions.

First, we define some basic terminology. Given an infinite sequence of integers $a_0 \geq 0, a_1 \geq 1, a_2 \geq 1, \ldots$ let $[a_0, a_1, \ldots, a_n]$ denote the finite continued fraction

$$[a_0, a_1, \ldots, a_n] = a_0 + \cfrac{1}{a_1 + \cfrac{1}{a_2 + \cfrac{1}{\ddots + \cfrac{1}{a_n}}}}. \tag{4.1}$$

You may wonder how we could possibly hope to combinatorially interpret a statement like $[2, 3, 4, 2] = \frac{67}{29}$ because the right side of the equation is not an integer. But since the numerator and denominator are integer-valued, we have every right to expect that the numbers 67 and 29 are somehow counting a problem that depends on $2, 3, 4$ and 2. For a given collection of integers, let p and q be functions producing the numerator and denominator of the resulting simplified continued fraction in lowest terms, i.e.,

$$[a_0, a_1, \ldots, a_n] = \frac{p(a_0, a_1, \ldots, a_n)}{q(a_0, a_1, \ldots, a_n)}.$$

For example, $p(2, 3, 4, 2) = 67$ and $q(2, 3, 4, 2) = 29$.

Naturally, since $[a] = \frac{a}{1}$, we have

$$p(a) = a \text{ and } q(a) = 1. \tag{4.2}$$

More complicated continued fractions can be computed recursively. By equation (4.1), for $n \geq 1$,

$$\begin{aligned}
[a_0, a_1, \ldots, a_n] &= a_0 + \frac{1}{[a_1, \ldots, a_n]} \\
&= a_0 + \frac{q(a_1, \ldots, a_n)}{p(a_1, \ldots, a_n)} \\
&= \frac{a_0 p(a_1, \ldots, a_n) + q(a_1, \ldots, a_n)}{p(a_1, \ldots, a_n)}.
\end{aligned}$$

Notice that the fraction on the right side must be in lowest terms since any number dividing the numerator and denominator must necessarily divide $p(a_1, \ldots, a_n)$ and $q(a_1, \ldots, a_n)$ which have no common factors. Thus,

$$p(a_0, a_1, \ldots, a_n) = a_0 p(a_1, \ldots, a_n) + q(a_1, \ldots, a_n), \tag{4.3}$$

$$q(a_0, a_1, \ldots, a_n) = p(a_1, \ldots, a_n). \tag{4.4}$$

Now let's do some combinatorics. For a sequence of numbers $a_0, a_1, \ldots, a_n$, consider the following tiling problem. Let $P(a_0, a_1, \ldots, a_n)$ count the ways to tile an $(n+1)$-board with dominoes and *stackable* square tiles. Nothing can be stacked on top of a domino, but for $0 \leq i \leq n$, the ith cell may be covered by a stack of as many as a_i square tiles. Figure 4.1 shows an untiled $(n + 1)$-board with the *height conditions* $a_0, a_1, \ldots, a_n$ indicated. Figure 4.2 gives an example of a valid tiling for a 12-board with height conditions 5, 10, 3, 1, 4, 8, 2, 7, 7, 4, 2, 3.

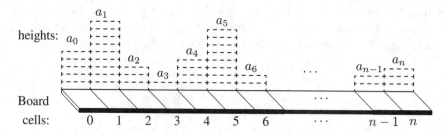

Figure 4.1. An empty $(n + 1)$-board.

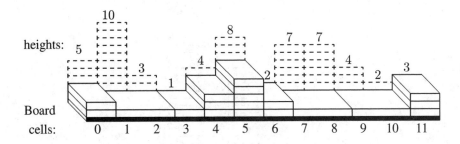

Figure 4.2. A tiling satisfying the height conditions 5, 10, 3, 1, 4, 8, 2, 7, 7, 4, 2, 3.

We define

$$Q(a_0, a_1, \ldots, a_n) = P(a_1, \ldots, a_n). \tag{4.5}$$

Thus $Q(a_0, a_1, \ldots, a_n)$ also counts the number of ways to tile an n-board with height conditions $a_1, \ldots, a_n$. (Note that the first cell has been removed.) Naturally, a 1-board with height condition a can be tiled a ways and an empty board can be tiled just one way. Thus,

$$P(a) = a \text{ and } Q(a) = 1. \tag{4.6}$$

Counting tilings of boards with two or more cells can be computed recursively by conditioning on how many squares cover the first cell or if a domino covers the first two cells. That is, for $n \geq 1$,

$$\begin{aligned} P(a_0, a_1, \ldots, a_n) &= a_0 P(a_1, \ldots, a_n) + P(a_2, \ldots, a_n) \\ &= a_0 P(a_1, \ldots, a_n) + Q(a_1, \ldots, a_n). \end{aligned} \tag{4.7}$$

By examining equations (4.2) through (4.7), we see that functions p and q satisfy the same initial conditions and recurrence relations as P and Q. Thus, we have

$$p(a_0, a_1, \ldots, a_n) = P(a_0, a_1, \ldots, a_n)$$

and

$$q(a_0, a_1, \ldots, a_n) = Q(a_0, a_1, \ldots, a_n).$$

Consequently, we have the following theorem.

Combinatorial Theorem 9 *Let $a_0, a_1, \ldots$ be a sequence of positive integers, and for $n \geq 0$, suppose the continued fraction $[a_0, a_1, \ldots, a_n]$ is equal to $\frac{p_n}{q_n}$, in lowest terms. Then for $n \geq 0$, p_n counts the ways to tile an $(n+1)$-board with height conditions $a_0, a_1, \ldots, a_n$ and q_n counts the ways to tile an n-board with height conditions $a_1, \ldots, a_n$.*

For example, the beginning of the "π-board" (see Figure 4.3) given by $[3, 7, 15]$ can be tiled 333 ways by either using all squares ($3 \times 7 \times 15 = 315$ ways), a stack of squares followed by a domino (three ways) or a domino followed by a stack of squares (fifteen ways). Removing the initial cell, the $[7, 15]$ board can be tiled 106 ways (105 ways for all squares, and one way for a single domino.) This produces the π approximation $[3, 7, 15] = \frac{333}{106}$. That is,

$$3 + \cfrac{1}{7 + \frac{1}{15}} = \frac{333}{106}.$$

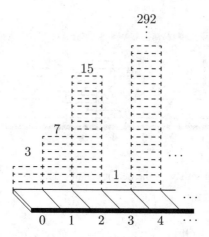

Figure 4.3. The beginning of the π board.

The curious reader might wonder what happens if we are allowed to stack dominoes in addition to stacking squares. This leads to a more general situation which will be explored in Section 4.3.

4.2   Identities

Armed with our tiling interpretation, many continued fraction identities practically reduce to "proofs without words". Traditionally, continued fractions are not computed by recurrences (4.3) and (4.4), but rather by the following relation.

Identity 106 *Let* $a_0 \geq 0, a_1 > 0, a_2 > 0, \ldots$, *and for* $n \geq 0$, *let* $[a_0, a_1, \ldots, a_n] = \frac{p_n}{q_n}$ *in lowest terms. Then*

  a) $p_0 = a_0$, $q_0 = 1$, $p_1 = a_0 a_1 + 1$, $q_1 = a_1$.
  b) For $n \geq 2$, $p_n = a_n p_{n-1} + p_{n-2}$.
  c) For $n \geq 2$, $q_n = a_n q_{n-1} + q_{n-2}$.

Part a) is easy to see both algebraically and combinatorially. Parts b) and c) however, are much easier to see combinatorially. We present the proof of b) only, since the proof of c) is virtually unchanged.

Question: For $n \geq 2$, in how many ways can the $(n+1)$-board with height conditions $a_0, a_1, \ldots, a_n$ be tiled by dominoes and stackable squares?

Answer 1: By Combinatorial Theorem 9, there are p_n such tilings.

Answer 2: Condition on the last tile. There are a_n ways for the tiling to end with a square, and the preceding board may be tiled p_{n-1} ways. There is one way to end with a domino, and the preceding board may be tiled p_{n-2} ways. Consequently, there are $a_n p_{n-1} + p_{n-2}$ such tilings.

As an immediate corollary of Combinatorial Theorem 9 or the previous identity, we have

Identity 107 *If* $a_i = 1$ *for all* $i \geq 0$, *then* $[a_0, a_1, \ldots, a_n] = f_{n+1}/f_n$.

The previous identity can be "extended":

Identity 108 *For all* $n \geq 1$, $[2, 1, 1, \ldots, 1, 1, 2] = f_{n+3}/f_{n+1}$, *where* $a_0 = 2$, $a_n = 2$, *and* $a_i = 1$ *for all* $0 < i < n$.

Denominator Set 1: The set of square-domino tilings of an n-board, where the last tile can be a domino, a square, or a stack of two squares. By Combinatorial Theorem 9, this set has size $p(1, 1, \ldots, 1, 1, 2) = q(2, 1, 1, \ldots, 1, 1, 2)$.

Denominator Set 2: The set of square-domino tilings of an $(n+1)$-board. This set has size f_{n+1}.

Correspondence: Let T be an $(n+1)$-tiling. If T ends with a square, then remove it to create an n-tiling with no stacked squares. If T ends with a domino, then "fold" that domino to create an n-tiling that ends with a stack of two squares.

Numerator Set 1: The set of square-domino tilings of an $(n+1)$-board, where the first or last tile can be a domino, a square, or a stack of two squares. By Combinatorial Theorem 9, this set has size $p(2, 1, 1, \ldots, 1, 1, 2)$.

Numerator Set 2: The set of square-domino tilings of an $(n+3)$-board. This set has size has f_{n+3}.

Correspondence: By applying the same "delete a square or fold a domino" procedure to the first and last tile, an $(n+3)$-tiling can be converted to an $(n+1)$-tiling that is allowed to have a stack of two squares at either end.

Other Fibonacci and Lucas identities are presented in the exercises, Identities 115–121. Next we prove the reversal identity mentioned at the top of the chapter. It is typically proved using an induction argument which we feel yields little insight. We hope you agree that the combinatorial proof is more satisfying.

Identity 109 *Suppose* $[a_0, a_1, \ldots, a_{n-1}, a_n] = p_n/q_n$. *Then for* $n \geq 1$, *we have*

$$[a_n, a_{n-1}, \ldots, a_1, a_0] = \frac{p_n}{p_{n-1}}.$$

Question (numerator): In how many ways can an $(n+1)$-board with height conditions $a_n, a_{n-1}, \ldots, a_1, a_0$ be tiled with dominoes and stackable squares?

Question (denominator): In how many ways can an n-board with height conditions $a_{n-1}, \ldots, a_1, a_0$ be tiled with dominoes and stackable squares?

Answer 1 (numerator and denominator): The answers are the numerator and denominator respectively of $[a_n, a_{n-1}, \ldots, a_1, a_0]$.

Answer 2 (numerator): There is a one-to-one correspondence between tilings that satisfy height conditions $a_n, a_{n-1}, \ldots, a_1, a_0$ and tilings that satisfy height conditions $a_0, a_1, \ldots, a_{n-1}, a_n$ by simply rotating the board 180 degrees. Hence the numerator generates p_n tilings.

Answer 2 (denominator): By the same one-to-one correspondence, there are as many tilings that satisfy the height conditions $a_{n-1}, \ldots, a_1, a_0$ as there are tilings that satisfy $a_0, a_1, \ldots, a_{n-1}$, namely p_{n-1}.

We define the *infinite* continued fraction $[a_0, a_1, a_2, \ldots]$ to be the limit of $[a_0, a_1, \ldots, a_n]$ as $n \to \infty$. As we shall see later, this limit always exists and is some irrational number α. The rational number $r_n = [a_0, a_1, \ldots, a_n] = p_n/q_n$ is called the nth *convergent* of α.

Let $\mathcal{P}_n$ and $\mathcal{Q}_n$ denote the set of all square-domino tilings with stackable square tiles over cells $0, \ldots, n$ and $1, \ldots, n$, respectively, with height conditions given by $a_0, \ldots, a_n$. Note that $|\mathcal{P}_n| = p_n$ and $|\mathcal{Q}_n| = q_n$.

The next few identities are useful for measuring the rate of convergence of convergents. The first one shows how convergents get closer to one another.

Identity 110 *The difference between consecutive convergents of $[a_0, a_1, \ldots]$ is:*

$$r_n - r_{n-1} = \frac{(-1)^{n-1}}{q_n q_{n-1}}.$$

Equivalently, after multiplying both sides by $q_n q_{n-1}$, we have

$$p_n q_{n-1} - p_{n-1} q_n = (-1)^{n-1}.$$

Set 1: The set $\mathcal{P}_n \times \mathcal{Q}_{n-1}$, which can be interpreted as the set of tilings of two boards, where the top board has cells $0, 1, \ldots, n$ with height conditions $a_0, a_1, \ldots, a_n$, and the bottom board has cells $1, \ldots, n-1$ with height conditions $a_1, \ldots, a_{n-1}$. This set has size $p_n q_{n-1}$.

Set 2: The set $\mathcal{P}_{n-1} \times \mathcal{Q}_n$, the set of tilings of two boards, where the top board has cells $0, 1, \ldots, n-1$ with height conditions $a_0, a_1, \ldots, a_{n-1}$, and the bottom board has cells $1, \ldots, n$ with height conditions $a_1, \ldots, a_n$. This set has size $p_{n-1} q_n$.

Correspondence: We exhibit an almost one-to-one correspondence between Set 1 and Set 2. Consider $(S, T) \in \mathcal{P}_n \times \mathcal{Q}_{n-1}$. As in previous chapters, for $i \geq 1$, we say (S, T) has a fault at cell i if both S and T have tiles that end at i. We say (S, T) has a fault at cell 0 if S has a square at cell 0. For instance, in Figure 4.4, there are faults at cells 0, 3, 5, and 6.

If (S, T) has a fault, construct (S', T') by swapping the "tails" of S and T after the rightmost fault. See Figure 4.5. Note that $(S', T') \in \mathcal{P}_{n-1} \times \mathcal{Q}_n$. Since (S', T') has the same rightmost fault as (S, T), this procedure is completely reversible.

Notice when either S or T contains a square, (S, T) must have a fault. Thus the only fault-free pairs occur when S and T consist of all dominoes in staggered

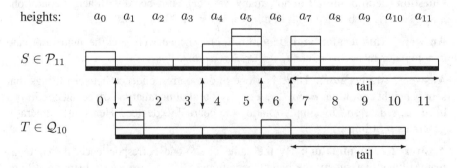

Figure 4.4. A pair of tilings with faults and tails indicated.

heights: a_0 a_1 a_2 a_3 a_4 a_5 a_6 a_7 a_8 a_9 a_{10} a_{11}

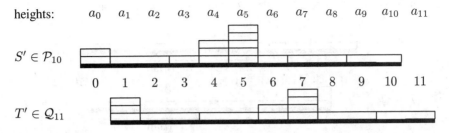

Figure 4.5. Result of swapping tails in Figure 4.4.

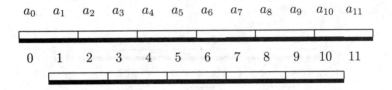

Figure 4.6. The fault-free pair consists of staggered dominoes.

formation as illustrated in Figure 4.6. When n is odd (i.e., S and T both cover an even number of cells), there is precisely one fault-free element of $\mathcal{P}_n \times \mathcal{Q}_{n-1}$ and no fault-free elements of $\mathcal{P}_{n-1} \times \mathcal{Q}_n$. Hence when n is odd, $|\mathcal{P}_n \times \mathcal{Q}_{n-1}| - |\mathcal{P}_{n-1} \times \mathcal{Q}_n| = 1$. Similarly when n is even, there are no fault-free elements of $\mathcal{P}_n \times \mathcal{Q}_{n-1}$ and exactly one fault-free element of $\mathcal{P}_{n-1} \times \mathcal{Q}_n$. Hence when n is even, $|\mathcal{P}_n \times \mathcal{Q}_{n-1}| - |\mathcal{P}_{n-1} \times \mathcal{Q}_n| = -1$.

Treating the odd and even case together, we obtain

$$p_n q_{n-1} - p_{n-1} q_n = (-1)^{n-1}.$$

Notice that the previous identity also implies that p_n/q_n is in lowest terms, since we have an integer combination of p_n and q_n producing ± 1. The next identity shows that the even convergents are increasing, while the odd convergents are decreasing.

Identity 111 $r_n - r_{n-2} = (-1)^n a_n/q_n q_{n-2}$. *Equivalently, after multiplying both sides by $q_n q_{n-2}$, we have*

$$p_n q_{n-2} - p_{n-2} q_n = (-1)^n a_n.$$

Set 1: $\mathcal{P}_n \times \mathcal{Q}_{n-2}$. This set has size $p_n q_{n-2}$.

Set 2: $\mathcal{P}_{n-2} \times \mathcal{Q}_n$. This set has size $p_{n-2} q_n$.

Correspondence: We use tail swapping to create an almost one-to-one correspondence between the sets $\mathcal{P}_n \times \mathcal{Q}_{n-2}$ and $\mathcal{P}_{n-2} \times \mathcal{Q}_n$. The proof is essentially given in Figures 4.7, 4.8, and 4.9. By tail swapping after the last fault, we have a one-to-one correspondence between the elements of $\mathcal{P}_n \times \mathcal{Q}_{n-2}$ and $\mathcal{P}_{n-2} \times \mathcal{Q}_n$ that have faults (Figures 4.7 and 4.8).

The only unmatched elements are those that are fault-free. When n is odd, there are no fault-free elements of $\mathcal{P}_n \times \mathcal{Q}_{n-2}$, but there are a_n fault-free elements of

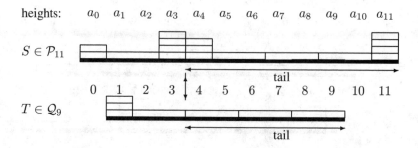

heights: a_0 a_1 a_2 a_3 a_4 a_5 a_6 a_7 a_8 a_9 a_{10} a_{11}

Figure 4.7. An element of $\mathcal{P}_{11} \times \mathcal{Q}_9$ with rightmost fault indicated.

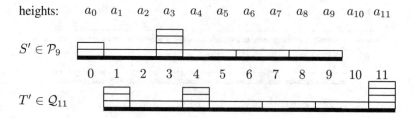

Figure 4.8. The result of swapping tails in Figure 4.7.

$\mathcal{P}_{n-2} \times \mathcal{Q}_n$, consisting of a stack of squares on the nth cell, and dominoes everywhere else (Figure 4.9).

Likewise when n is even, there are no fault-free elements of $\mathcal{P}_{n-2} \times \mathcal{Q}_n$, but there are precisely a_n fault-free elements of $\mathcal{P}_n \times \mathcal{Q}_{n-2}$, consisting of a stack of squares on the nth cell, and dominoes everywhere else. Thus we have established $|\mathcal{P}_n \times \mathcal{Q}_{n-2}| - |\mathcal{P}_{n-2} \times \mathcal{Q}_n| = (-1)^n a_n$, as desired.

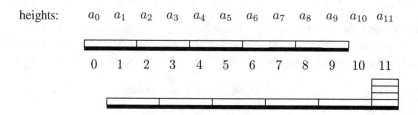

Figure 4.9. Problem pairings are fault-free.

Using the combinatorially clear fact that $q_n \to \infty$ as $n \to \infty$, the last two identities demonstrate that $[r_0, r_1], [r_2, r_3], [r_4, r_5], \ldots$ is a sequence of nested intervals whose lengths are going to zero. Hence, $\lim_{n \to \infty} r_n$ exists.

If we let $r = \lim_{n \to \infty} r_n$, we see by our nested intervals and Identity 110 that

$$0 < |r - r_n| < |r_{n+1} - r_n| < \frac{1}{q_{n+1} q_n} < \frac{1}{q_n^2}.$$

Hence

$$0 < \left| r - \frac{p_n}{q_n} \right| < \frac{1}{q_n^2}.$$

It would be a crime to go this far and not prove that an infinite continued fraction must be an irrational number. Let $r = [a_0, a_1, a_2, \ldots]$. Multiplying the last inequality by q_n gives us

$$0 < |rq_n - p_n| < \frac{1}{q_n}.$$

Now if $r = \frac{a}{b}$, then multiplying by $b > 0$ implies that for *all* $n \geq 0$,

$$0 < |aq_n - bp_n| < \frac{b}{q_n}.$$

But the middle quantity is obviously an integer and the right quantity gets arbitrarily small as n gets large. Since there are no integers between 0 and 1, we have reached a contradiction and must conclude that r is irrational.

Continuants

Next we examine, for $i \leq j$, the quantity $K(i, j)$ which counts the number of tilings of the sub-board with cells $i, i+1, \ldots, j$ with height conditions $a_i, a_{i+1}, \ldots, a_j$. We see that $K(i, j)$ is the numerator of the finite continued fraction $[a_i, , \ldots, a_j]$ and the denominator of the finite continued fraction $[a_{i-1}, \ldots, a_j]$. Also, we define $K(j + 1, j) = 1$. The $K(i, j)$ are identical to the classical *continuants* of Euler [40].

The following theorem, due to Euler, can also be proved by the now familiar tail swapping technique.

Identity 112 *For $i < m < j < n$,*

$$K(i, j)K(m, n) - K(i, n)K(m, j) = (-1)^{j-m}K(i, m - 2)K(j + 2, n).$$

This result follows by considering tilings of sub-boards S from cells i to j and T from m to n. Every faulty pair (S, T) corresponds to another faulty pair (S', T') obtained by swapping the tails after the last fault. The term on the right side of Identity 112 counts the number of fault-free tilings that only occur when the overlapping regions (of S and T, or of S' and T', depending on the parity of $j - m$) consist entirely of dominoes in staggered formation. See Figures 4.10 and 4.11. Setting $i = 0$ and $m = 1$, Identity 112 generalizes Identities 110 and 111 by allowing us to compare arbitrary convergents r_j and r_n.

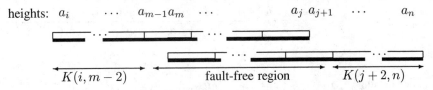

Figure 4.10. When $j - m$ is even, there are $K(i, m - 2)K(j + 2, n)$ fault-free tilings (S, T).

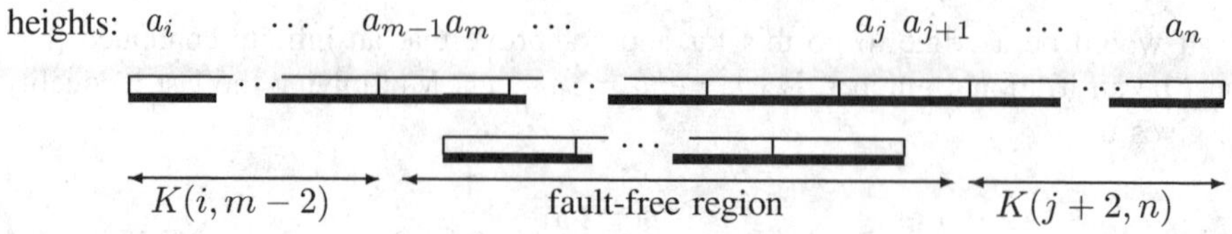

heights: a_i $\cdots$ $a_{m-1} a_m$ $\cdots$ $a_j\ a_{j+1}$ $\cdots$ a_n

$$\underbrace{}_{K(i,\,m-2)} \quad \underbrace{}_{\text{fault-free region}} \quad \underbrace{}_{K(j+2,\,n)}$$

Figure 4.11. When $j - m$ is odd, there are $K(i, m - 2)K(j + 2, n)$ fault-free tilings (S', T').

4.3   Nonsimple Continued Fractions

Finally, we examine continued fractions of the form

$$a_0 + \cfrac{b_1}{a_1 + \cfrac{b_2}{a_2 + \cfrac{b_3}{\ddots + \cfrac{b_n}{a_n + \ddots}}}} \tag{4.8}$$

where for $i > 0$, a_i and b_i are positive integers, and a_0 is a nonnegative integer. We shall denote the *nonsimple finite continued fractions* by

$$[a_0, (b_1, a_1), (b_2, a_2), \dots, (b_n, a_n)] = a_0 + \cfrac{b_1}{a_1 + \cfrac{b_2}{a_2 + \cfrac{b_3}{\ddots + \cfrac{b_n}{a_n}}}}. \tag{4.9}$$

As before, we shall let p and q be functions that produce the numerator and denominator of a finite continued fraction. When evaluating a simple finite continued fraction from the "bottom up", the result is always a fraction in lowest terms, but this is not necessarily the case for nonsimple ones. For example $[3, (2, 4)] = \frac{14}{4}$. We shall write our numerator and denominator without reducing to lowest terms. Hence $p[3, (2, 4)] = 14$ and $q[3, (2, 4)] = 4$. Functions p and q still satisfy the initial conditions $p[a] = a$, $q[a] = 1$, and for $n \geq 1$,

$$[a_0, (b_1, a_1), \dots, (b_n, a_n)]$$

$$= a_0 + \frac{b_1}{[a_1, (b_2, a_2), \dots, (b_n, a_n)]}$$

$$= a_0 + \frac{b_1 q[a_1, (b_2, a_2), \dots, (b_n, a_n)]}{p[a_1, (b_2, a_2), \dots, (b_n, a_n)]}$$

$$= \frac{a_0 p[a_1, (b_2, a_2), \dots, (b_n, a_n)] + b_1 q[a_1, (b_2, a_2), \dots, (b_n, a_n)]}{p[a_1, (b_2, a_2), \dots, (b_n, a_n)]}.$$

Thus,

$$p[a_0, (b_1, a_1), \dots, (b_n, a_n)] = a_0 p[a_1, (b_2, a_2), \dots, (b_n, a_n)]$$
$$+ b_1 q[a_1, (b_2, a_2), \dots, (b_n, a_n)],$$
$$q[a_0, (b_1, a_1), \dots, (b_n, a_n)] = p[a_1, (b_2, a_2), \dots, (b_n, a_n)].$$

Now we explore a related tiling problem. Suppose we allow dominoes to be stacked as well as squares. Specifically, suppose for $i \geq 1$, we impose height conditions $b_1, b_2, \ldots$ so that we may stack as many as b_i dominoes on cells $i - 1$ and i. We let $P[(a_0, (b_1, a_1), \ldots, (b_n, a_n)]$ count the number of ways to tile an $(n + 1)$-board with cells $0, 1, \ldots, n$ and height conditions $a_0, \ldots, a_n$ and $b_1 \ldots, b_n$ for the squares and dominoes respectively. We let Q_n count the same problem with cell 0 removed (along with height conditions a_0 and b_1). Thus, $P[a] = a$, $Q[a] = 1$, and for $n \geq 1$,

$$Q[(a_0, (b_1, a_1), \ldots, (b_n, a_n)] = P[a_1, (b_2, a_2), \ldots, (b_n, a_n)].$$

By conditioning on the first tile, we see that P satisfies

$$P[a_0, (b_1, a_1), \ldots, (b_n, a_n)] = a_0 P[a_1, (b_2, a_2), \ldots, (b_n, a_n)]$$
$$+ b_1 Q[a_1, (b_2, a_2), \ldots, (b_n, a_n)]$$

Consequently, we have the following theorem.

Combinatorial Theorem 10 *Let $a_0, a_1, \ldots$ be a sequence of positive integers. For $n \geq 1$, suppose the continued fraction $[a_0, (b_1, a_1), \ldots, (b_n, a_n)]$ computed by recurrence (4.9) is equal to $\frac{p_n}{q_n}$. Then for $n \geq 0$, p_n counts the ways to tile an $(n + 1)$-board with height conditions $a_0, (b_1, a_1), \ldots, (b_n, a_n)$ and q_n counts the ways to tile an n-board with height conditions $a_1, (b_2, a_2) \ldots, (b_n, a_n)$.*

Some consequences of this theorem are explored in the exercises.

4.4 Get Real Again!

Can we make sense out of $[a_0, a_1, \ldots, a_n]$ when some of the a_is are real or complex numbers? Essentially, yes! Just as we did at the end of the last chapter, instead of giving a square at cell i a_i choices, we give that square a *weight* of a_i, dominoes are given a weight of 1, and the weight of the tiling is the *product* of the weights of its tiles. Then we can define $P(a_0, a_1, \ldots, a_n)$ and $Q(a_0, a_1, \ldots, a_n)$ as the sum of the weights of all square-domino tilings over cells $0, 1, \ldots, n$ and cells $1 \ldots, n$, respectively. The continued fraction for $[a_0, a_1, \ldots, a_n]$ will simplify to $P(a_0, a_1, \ldots, a_n)/Q(a_0, a_1, \ldots, a_n)$ provided that no division by zero takes place along the way. Nonsimple continued fractions can be handled in this way too when evaluating $[a_0, (a_1, b_1), \ldots, (a_n, b_n)]$ by assigning a weight of b_i to dominoes that cover cells $i - 1$ and i.

4.5 Notes

We thank Christopher Hanusa who, as an undergraduate, mentioned that some of the Gibonacci identities reminded him of continued fraction identities. Some of the material in this chapter originally appeared in [12], and we are grateful to Jim Propp for suggesting to us a derivation of Combinatorial Theorem 9 that does not rely on Identity 106. Ira Gessel suggested many of the exercises that follow. For more information on continued fractions (although not from a combinatorial perspective) see [31] or [39].

4.6   Exercises

1. Directly count the tilings of the board with height conditions $3, 7, 15, 1$ to show that $[3, 7, 15, 1] = \frac{355}{113}$.

2. Find (nonsimple) continued fractions that produce numerators for Lucas numbers and Gibonacci Numbers.

Prove each of the identities below by a direct combinatorial argument.

Identity 113 *For* $n \geq 0$, $[a_0, a_1, \ldots, a_n, 2] = [a_0, a_1, \ldots, a_n, 1, 1]$.

Identity 114 *For* $n \geq 0$, *if* $m \geq 2$ *then*

$$[a_0, a_1, \ldots, a_n, m] = [a_0, a_1, \ldots, a_n, m - 1, 1].$$

Identity 115 *For* $n \geq 0$, $[3, 1, 1, \ldots, 1] = L_{n+2}/f_n$, *where* $a_0 = 3$, *and* $a_i = 1$ *for all* $0 < i \leq n$.

Identity 116 *For* $n \geq 1$, $[1, 1, \ldots, 1, 3] = L_{n+2}/L_{n+1}$, *where* $a_n = 3$, *and* $a_i = 1$ *for all* $0 \leq i < n$.

Identity 117 *For* $n \geq 1$, $[4, 4, \ldots, 4, 3] = f_{3n+3}/f_{3n}$, *where* $a_n = 3$, *and* $a_i = 4$ *for all* $0 \leq i < n$.

Identity 118 *For* $n \geq 1$, $[4, 4, \ldots, 4, 5] = f_{3n+4}/f_{3n+1}$, *where* $a_n = 5$, *and* $a_i = 4$ *for all* $0 \leq i < n$.

Identity 119 *Let* $a_i = 4$ *for* $0 \leq i \leq n$. *Then* $[4, 4, \ldots, 4] = f_{3n+5}/f_{3n+2}$.

Identity 120 *For* $n \geq 1$, $[2, 4, \ldots, 4, 3] = L_{3n+1}/f_{3n}$, *where* $a_0 = 2$, $a_n = 3$, *and* $a_i = 4$ *for all* $0 < i < n$.

Identity 121 *For* $n \geq 1$, $[2, 4, \ldots, 4, 5] = L_{3n+2}/f_{3n+1}$, *where* $a_0 = 2$, $a_n = 5$, *and* $a_i = 4$ *for all* $0 < i < n$.

Identity 122 *For nonsimple continued fractions,*

$$P_n = a_n P_{n-1} + b_n P_{n-2},$$
$$Q_n = a_n Q_{n-1} + b_n Q_{n-2}$$

for $n \geq 2$, *with initial conditions* $P_0 = a_0$, $P_1 = a_1 a_0 + b_1$, $Q_0 = 1$, $Q_1 = a_1$.

Identity 123 *For nonnegative integers* s, t, *let* $u_0 = 1$, $u_1 = s$, *and for* $n \geq 2$, *define* $u_n = s u_{n-1} + t u_{n-2}$. *Then the nonsimple continued fraction*

$$[a_0, (b_1, a_1), (b_2, a_2), \ldots, (b_n, a_n)] = [s, (t, s), (t, s), \ldots, (t, s)] = u_{n+1}/u_n.$$

Identity 124 *For nonnegative integers* s, t, *let* $v_0 = 2$, $v_1 = s$, *and for* $n \geq 2$, *define* $v_n = s v_{n-1} + t v_{n-2}$. *Then the nonsimple continued fraction*

$$[a_0, (b_1, a_1), (b_2, a_2), \ldots, (b_{n-1}, a_{n-1}), (b_n, a_n)]$$
$$= [s, (t, s), (t, s), \ldots, (t, s), (2t, s)] = v_{n+1}/v_n.$$

Uncounted Identities

The identities listed below are in need of combinatorial proof.

1. Combinatorially prove:

$$\frac{F_{(t+1)m}}{F_{tm}} = L_m - \cfrac{(-1)^m}{L_m - \cfrac{(-1)^m}{L_m - \cfrac{(-1)^m}{\ddots - \cfrac{(-1)^m}{L_m}}}}, \qquad (4.10)$$

where the number L_m appears t times. Note that this is a generalization of our formulas for $[1, 1, \ldots, 1]$ and $[4, 4, \ldots, 4]$.

2. Euler proved

$$e = 2 + \cfrac{1}{1 + \cfrac{1}{2 + \cfrac{2}{3 + \cfrac{3}{4 + \cfrac{4}{5 + \ddots}}}}}. \qquad (4.11)$$

What are the combinatorial implications for e?

CHAPTER 5

Binomial Identities

Definition The *binomial coefficient* $\binom{n}{k}$ is the number of k-element subsets of $\{1,\dots,n\}$.

Definition The *multichoose coefficient* $\left(\binom{n}{k}\right)$ is the number of k-element multisubsets of $\{1,\dots,n\}$.

Examples of binomial coefficients are $\binom{4}{0}=1$, $\binom{4}{1}=4$, $\binom{4}{2}=6$, $\binom{4}{3}=4$, and $\binom{4}{4}=1$.
Examples of multichoose coefficients are $\left(\binom{4}{0}\right)=1$, $\left(\binom{4}{1}\right)=4$, $\left(\binom{4}{2}\right)=10$, $\left(\binom{4}{3}\right)=20$, and $\left(\binom{4}{4}\right)=35$.

5.1 Combinatorial Interpretations of Binomial Coefficients

Binomial coefficients were born to count! Unlike most of the quantities we have discussed in this book, binomial coefficients are almost always defined as the answer to a counting problem. Specifically, we define $\binom{n}{k}$ to be the number of k-element subsets of $\{1,2,\dots,n\}$. Put another way, $\binom{n}{k}$ counts the ways to select a committee of k students from a class of n students where the order of the selection is not important. By definition we have, for $n \geq 0$, $\binom{n}{0}=1$, and for $k < 0$, $\binom{n}{k}=0$. (Although it's possible to define $\binom{n}{k}$ for negative values of n, we will not do so here.)

Binomial coefficients have a simple algebraic formula

$$\binom{n}{k} = \frac{n!}{k!(n-k)!} \tag{5.1}$$

which can be easily seen by the following identity:

Identity 125 *For* $0 \leq k \leq n$, $n! = \binom{n}{k}k!(n-k)!$

Question: How many ways can the numbers 1 through n be arranged in a list?

Answer 1: There are $n!$ arrangements since the first number can be chosen n ways, the next number can be chosen $n-1$ ways, and so on. (We shall have more to say about $n!$ in Chapter 7.)

Answer 2: Condition on which numbers are among the first k in our arrangement. There are, by definition, $\binom{n}{k}$ ways to choose which of the n numbers appear among the first k. Once these are chosen, there are $k!$ ways to arrange them, followed by $(n - k)!$ ways to arrange the remaining elements. Hence the numbers 1 through n can be arranged in $\binom{n}{k}k!(n - k)!$ ways.

We shall take pains to avoid invoking equation (5.1), in the same way that we avoided using Binet's formula (Identity 240) when proving identities for Fibonacci numbers in Chapter 1. Our goal is to understand binomial identities entirely from their combinatorial definition and to avoid algebraic arguments (such as proofs by induction) as much as possible.

5.2   Elementary Identities

In this section, we present simple combinatorial proofs of binomial coefficient identities. Although the arguments we present in this section are quite well-known, they are beautiful nonetheless. In subsequent sections of this chapter, the arguments will become trickier.

Identity 126 *For* $0 \leq k \leq n$,

$$\binom{n}{k} = \binom{n}{n - k}.$$

Question: How many ways can we create a size k committee of students from a class of n students?

Answer 1: By definition, $\binom{n}{k}$.

Answer 2: We may choose $n - k$ students to exclude from the committee, which can be done $\binom{n}{n-k}$ ways.

Identity 127 *For* $0 \leq k \leq n$, *(except* $n = k = 0$),

$$\binom{n}{k} = \binom{n - 1}{k} + \binom{n - 1}{k - 1}.$$

Question: How many ways can we create a size k committee of students from a class of n students?

Answer 1: As before, $\binom{n}{k}$.

Answer 2: Condition on whether or not student n is on the committee. There are $\binom{n-1}{k}$ committees that exclude student n, and $\binom{n-1}{k-1}$ committees that include student n.

Identity 127 (along with initial conditions $\binom{0}{0} = 1$ and $\binom{n}{k} = 0$ for $n < k$) can be used to generate binomial coefficients in a convenient table known as Pascal's Triangle. See Figure 5.1.

Although the previous identities are easy to prove by using the algebraic formula for $\binom{n}{k}$ given in Identity 125, the next identity is not at all obvious from the factorial definition of $\binom{n}{k}$. Note that the sum on the left is finite, since $\binom{n}{k} = 0$ for $k > n$.

Identity 128 *For* $n \geq 0$,

$$\sum_{k \geq 0} \binom{n}{k} = 2^n.$$

$n \backslash k$	0	1	2	3	4	5	6	7	8	9	10
0	1										
1	1	1									
2	1	2	1								
3	1	3	3	1							
4	1	4	6	4	1						
5	1	5	10	10	5	1					
6	1	6	15	20	15	6	1				
7	1	7	21	35	35	21	7	1			
8	1	8	28	56	70	56	28	8	1		
9	1	9	36	84	126	126	84	36	9	1	
10	1	10	45	120	210	252	210	120	45	10	1

Figure 5.1. Numbering our rows and columns with nonnegative integers, the number in row n and column k is $\binom{n}{k}$, and all missing entries are zero.

Question: How many ways can we create a committee (of any size) from a class of n students?

Answer 1: Since for $0 \le k \le n$, there are $\binom{n}{k}$ committees of size k, there are $\sum_{k \ge 0} \binom{n}{k}$ such committees.

Answer 2: Decide, student by student, whether or not to put that student on the committee. Since there are two possibilities for each student (on or off), there are 2^n possible committees.

Identity 129 *For $n \ge 1$,*

$$\sum_{k \ge 0} \binom{n}{2k} = 2^{n-1}.$$

Question: How many ways can we create a committee with an even number of members from a class of n students?

Answer 1: Since for $0 \le 2k \le n$, there are $\binom{n}{2k}$ committees of size $2k$, there are $\sum_{k \ge 0} \binom{n}{2k}$ such committees.

Answer 2: The first $n-1$ students can be freely chosen to be on or off of the committee, as in the previous proof. Once these choices are made, then the fate of the nth student is completely determined so that the final committee size is an even number. Consequently, there are 2^{n-1} such committees.

Notice that the last two identities imply that exactly half of all subsets of $\{1, \dots, n\}$ are even. Consequently, half of them must also be odd. Equivalently, this says

$$\sum_{k=0}^{n} \binom{n}{k} (-1)^k = 0.$$

We shall have more to say about such alternating sums in the next chapter.

Identity 130 *For $0 \le k \le n$,*

$$k \binom{n}{k} = n \binom{n-1}{k-1}.$$

Question: How many ways can we create a size k committee of students from a class of n students, where one of the committee members is designated as chair?

Answer 1: There are $\binom{n}{k}$ ways to choose the committee, then k ways to select the chair. Hence there are $k\binom{n}{k}$ possible outcomes.

Answer 2: First select the chair from the class of n students. Then from the remaining $n-1$ students, pick the remaining $k-1$ committee members. This can be done $n\binom{n-1}{k-1}$ ways.

The next identity can be treated as a continuation of Identity 130.

Identity 131 *For $n \geq 1$,*

$$\sum_{k=0}^{n} k\binom{n}{k} = n2^{n-1}.$$

Question: How many ways can we create a committee (of any size) from a class of n students, where one of the committee members is designated as chair?

Answer 1: For a committee of size k, where $0 \leq k \leq n$, there are $k\binom{n}{k}$ such committees. Altogether, we have $\sum_{k=0}^{n} k\binom{n}{k}$ possible outcomes.

Answer 2: First select the chair from the class of n students. Then from the remaining $n-1$ students, there are 2^{n-1} ways to choose a subset of them to form the rest of the committee.

Dividing both sides of the last identity by 2^n allows us to give a different combinatorial proof of the equivalent identity:

$$\frac{\sum_{k=0}^{n} k\binom{n}{k}}{2^n} = \frac{n}{2}.$$

Question: What is the average size of a subset of $\{1, 2, \ldots, n\}$?

Answer 1: We add up the sizes of all subsets and divide by the total number of subsets. Since for $0 \leq k \leq n$, there are $\binom{n}{k}$ subsets of size k, and there are 2^n subsets altogether, the average subset size is $\frac{\sum_{k=0}^{n} k\binom{n}{k}}{2^n}$.

Answer 2: Pair up each subset with its complement. Since each such pair has n elements, each complementary pair has an average of $\frac{n}{2}$ elements. Hence the average subset size is $\frac{n}{2}$.

The next identity, *Vandermonde's Identity*, has a simple combinatorial interpretation.

Identity 132 *For $m \geq 0$, $n \geq 0$,*

$$\binom{m+n}{k} = \sum_{j=0}^{k} \binom{m}{j}\binom{n}{k-j}.$$

Question: From a class of $m+n$ students, consisting of m men and n women, how many ways can one form a size k committee?

Answer 1: By definition, $\binom{m+n}{k}$.

Answer 2: Condition on the number of men on the committee. For $0 \leq j \leq k$, we can form a committee with j men by first choosing the men ($\binom{m}{j}$ ways), then the remaining $k - j$ committee members can be chosen from the women in $\binom{n}{k-j}$ ways. Altogether, there are $\sum_{j=0}^{k} \binom{m}{j}\binom{n}{k-j}$ such committees.

Many of the previous identities can be proved using algebraic methods based on the *Binomial Theorem*, but even that can be proved combinatorially.

Identity 133 *For $n \geq 0$,*

$$(x + y)^n = \sum_{k=0}^{n} \binom{n}{k} x^k y^{n-k}.$$

Question: In a class of n students, each student is given the choice of solving either one of x different algebra problems or one of y different geometry problems. How many different outcomes are possible?

Answer 1: Since each student has $x + y$ choices for which problem to solve, there are $(x + y)^n$ possible outcomes.

Answer 2: Condition on the number of students who choose to solve an algebra problem. For $0 \leq k \leq n$, there are $\binom{n}{k}$ ways to determine which k students chose to do an algebra problem, then x^k ways for them to decide which algebra problems to do, then y^{n-k} ways for the remaining $n - k$ students to decide which geometry problems to do. Altogether, there are $\sum_{k=0}^{n} \binom{n}{k} x^k y^{n-k}$ possible outcomes.

The proof above assumes that x and y are integers, although the theorem is true for real or complex values of x and y as well. There are several combinatorial ways around this issue. One way is to observe for any fixed y, both sides of the identity are degree n polynomials in x that agree on an infinite number of points. Hence they must be equal.

Another (slightly more algebraic) way to view this identity is to think of the expression

$$(x + y)^n = (x + y)(x + y) \cdots (x + y) \qquad (n \text{ times}),$$

and ask, "How many ways can one create an $x^k y^{n-k}$ term?" Each such term arises by choosing an x term from k of the $x + y$ factors, which can be done $\binom{n}{k}$ ways.

The next identity has an interesting application to number theory.

Identity 134 *For $0 \leq m \leq k \leq n$,*

$$\binom{n}{k}\binom{k}{m} = \binom{n}{m}\binom{n-m}{k-m}.$$

Question: In a class of n students, how many ways can we choose a size k committee that contains a size m subcommittee?

Answer 1: The committee can be chosen $\binom{n}{k}$ ways, then the subcommittee can be chosen $\binom{k}{m}$ ways.

Answer 2: First choose the m students who will be on the committee and the subcommittee. This can be done $\binom{n}{m}$ ways. From the remaining $n - m$ students, the $k - m$ students to be on the committee but not the subcommittee can be chosen $\binom{n-m}{k-m}$ ways.

As a simple consequence of this last identity, Erdős and Szekeres proved the following simple fact about binomial coefficients. (It seems that this was not known prior to 1978!)

Corollary 7 *For* $0 < m \le k < n$, $\binom{n}{m}$ *and* $\binom{n}{k}$ *have a nontrivial common factor. That is,* $\gcd(\binom{n}{m}, \binom{n}{k}) > 1$.

Proof. Suppose, to the contrary, that $\binom{n}{m}$ and $\binom{n}{k}$ are relatively prime. By Identity 134, $\binom{n}{m}$ divides $\binom{n}{k}\binom{k}{m}$. But since $\binom{n}{m}$ and $\binom{n}{k}$ have no common factors, it follows that $\binom{n}{m}$ divides $\binom{k}{m}$. This is impossible, since it is (combinatorially) clear that $\binom{n}{m}$ is greater than $\binom{k}{m}$.   ◇

5.3 More Binomial Coefficient Identities

For the identities in this section, it is more convenient to talk about subsets than committees. While Identity 128 proved that $\sum_{k=0}^{n} \binom{n}{k} = 2^n$, no general closed form exists for the partial sum $\sum_{k=0}^{m} \binom{n}{k}$ where $m < n$. However, if we interchange the roles of the fixed and the indexed variable in the binomial summation, a closed form for the partial sum does exist. Specifically:

Identity 135 *For* $0 \le k \le n$,

$$\sum_{m=k}^{n} \binom{m}{k} = \binom{n+1}{k+1}.$$

Question: How many $(k+1)$-subsets are contained in the set $\{1, 2, \ldots, n+1\}$?

Answer 1: By definition, $\binom{n+1}{k+1}$.

Answer 2: Condition on the largest number in the subset. A size $k+1$ subset with maximum element $m+1$ can be created $\binom{m}{k}$ ways. Since $m+1$ can be as small as $k+1$ and as large as $n+1$, there are $\binom{k}{k} + \binom{k+1}{k} + \cdots + \binom{n}{k}$ subsets in total.

Identity 136 *For* $0 \le k \le n/2$,

$$\sum_{m=k}^{n-k} \binom{m}{k}\binom{n-m}{k} = \binom{n+1}{2k+1}.$$

Question: How many $(2k+1)$-subsets are contained in the set $\{1, \ldots, n+1\}$?

Answer 1: By definition, $\binom{n+1}{2k+1}$.

Answer 2: Condition on the median number in the subset. In a size $2k+1$ subset, the median element will be the $(k+1)$st smallest element, with k elements below it and k elements above it. (For example, in the set $\{2, 3, 5, 8, 13\}$, the median element is 5.) Hence, the number of size $2k+1$ subsets with median element $m+1$ is $\binom{m}{k}\binom{n-m}{k}$. Since $m+1$ can range from $k+1$ to $n+1-k$, the identity follows.

By conditioning on the rth element of the set, we obtain the following generalization.

Identity 137 *For* $1 \leq r \leq k$,

$$\sum_{j=r}^{n+r-k} \binom{j-1}{r-1}\binom{n-j}{k-r} = \binom{n}{k}.$$

As we have seen before, binomial coefficients and Fibonacci numbers can't help running into each other. The next few identities are variations on the same theme.

Identity 138 *For* $t \geq 1, n \geq 0$,

$$\sum_{x_1 \geq 0} \sum_{x_2 \geq 0} \cdots \sum_{x_t \geq 0} \binom{n}{x_1}\binom{n-x_1}{x_2}\binom{n-x_2}{x_3}\cdots\binom{n-x_{t-1}}{x_t} = f_{t+1}^n.$$

Question: In how many ways can you create subsets $S_1, S_2, \ldots, S_t$, where $S_1 \subseteq \{1, 2, \ldots, n\}$ and for $2 \leq i \leq t$, $S_i \subseteq \{1, 2, \ldots, n\}$ and S_i is disjoint from S_{i-1}?

Answer 1: Condition on the size of each subset S_i. To create subsets that are "consecutively disjoint" with sizes $x_i = |S_i|$, $1 \leq i \leq n$, there are $\binom{n}{x_1}$ ways to create S_1. Then, since S_2 is disjoint from S_1, there are $\binom{n-x_1}{x_2}$ ways to create S_2. Since S_3 is disjoint from S_2, there are $\binom{n-x_2}{x_3}$ ways to create S_3 and so on. Thus there are $\binom{n}{x_1}\binom{n-x_1}{x_2}\binom{n-x_2}{x_3}\cdots\binom{n-x_{t-1}}{x_t}$ ways to create $S_1, \ldots, S_t$ with respective sizes $x_1, \ldots, x_t$. Altogether $S_1, S_2, \ldots, S_t$ can be created in

$$\sum_{x_1 \geq 0} \sum_{x_2 \geq 0} \cdots \sum_{x_t \geq 0} \binom{n}{x_1}\binom{n-x_1}{x_2}\binom{n-x_2}{x_3}\cdots\binom{n-x_{t-1}}{x_t}$$

ways.

Answer 2: For each element $j \in \{1, \ldots, n\}$, decide which subsets contain j. By construction, the subsets containing j must be nonconsecutive. Exercise 1 in Chapter 1 shows that there are f_{t+1} ways to select the nonconsecutive subsets containing j, among the sets $S_1, \ldots, S_t$. Hence the elements 1 through n can be placed into subsets in f_{t+1}^n ways.

For those that prefer the tiling approach from Chapter 1, here is a different proof of Identity 138.

Question: In how many ways can you create n square-domino tilings $T_1, \ldots, T_n$, each of length $t + 1$?

Answer 1: Each tiling can be created f_{t+1} ways, so there are f_{t+1}^n such tilings.

Answer 2: For each cell j, $1 \leq j \leq t$, let x_j denote the number of tilings that have a domino beginning at cell j. Conditioning on all possible values of $x_1, \ldots, x_t$, we have $\binom{n}{x_1}$ ways to decide which of $T_1, \ldots, T_n$ begin with a domino. (The rest begin with a square.) Among the $n - x_1$ tilings that do not begin with a domino, there are $\binom{n-x_1}{x_2}$ ways to choose which tilings have a domino beginning at cell 2. (Among these $n - x_1$ tilings, the unchosen ones have a square at cell 2.) Among the $n - x_2$ tilings that do not have a domino covering cells 2 and 3, there are $\binom{n-x_2}{x_3}$ ways to choose which tilings have a domino beginning at cell 3. Continuing in this fashion, $T_1, \ldots, T_n$ can be tiled in $\binom{n}{x_1}\binom{n-x_1}{x_2}\binom{n-x_2}{x_3}\cdots\binom{n-x_{t-1}}{x_t}$ ways, as desired.

Generalizing the previous argument, we obtain

Identity 139 *For* $t \geq 1, n \geq 0, c \geq 0$,

$$\sum_{x_1 \geq 0} \sum_{x_2 \geq 0} \cdots \sum_{x_t \geq 0} \binom{n-c}{x_1} \binom{n-x_1}{x_2} \binom{n-x_2}{x_3} \cdots \binom{n-x_{t-1}}{x_t} = f_t^c f_{t+1}^{n-c}.$$

Question: In how many ways can you create n square-domino tilings $T_1, \ldots, T_n$ of length $t+1$, where $T_1, \ldots, T_c$ must begin with a square?

Answer 1: There are $f_t^c f_{t+1}^{n-c}$ such tilings, since the first c t-tilings can be created f_t ways, and the remaining $n - c$ $(t+1)$-tilings can be created f_{t+1}^{n-c} ways.

Answer 2: The exact same reasoning as in the last proof applies here. The only difference is that the x_1 tilings that begin with dominoes must be chosen from $T_{c+1}, \ldots, T_n$. Hence the first step can be performed $\binom{n-c}{x_1}$ ways instead of $\binom{n}{x_1}$.

Identity 138 can be generalized in a different direction to produce a Lucas identity.

Identity 140 *For* $t \geq 1, n \geq 0$,

$$\sum_{x_1 \geq 0} \sum_{x_2 \geq 0} \cdots \sum_{x_t \geq 0} \binom{n}{x_1} \binom{n-x_1}{x_2} \binom{n-x_2}{x_3} \cdots \binom{n-x_{t-1}}{x_t} 2^{x_1} = L_{t+1}^n.$$

The proof is the same as in Identity 138, but now each of the x_1 tilings that begin with a domino is given one of two phases. Even more generally, we have

Identity 141 *For* $t \geq 1, n \geq 0$,

$$\sum_{x_1 \geq 0} \sum_{x_2 \geq 0} \cdots \sum_{x_t \geq 0} \binom{n}{x_1} \binom{n-x_1}{x_2} \binom{n-x_2}{x_3} \cdots \binom{n-x_{t-1}}{x_t} G_0^{x_1} G_1^{n-x_1} = G_{t+1}^n,$$

where G_j is the jth element of the Gibonacci sequence beginning with G_0 and G_1.

We remark that Identity 138 and its generalizations arose from our attempts to combinatorially prove the following generalization of Identity 5 from Chapter 1.

Identity 142 *For* $t \geq 1, n \geq 0$,

$$\sum_{x_1 \geq 0} \sum_{x_2 \geq 0} \cdots \sum_{x_t \geq 0} \binom{n-x_t}{x_1} \binom{n-x_1}{x_2} \binom{n-x_2}{x_3} \cdots \binom{n-x_{t-1}}{x_t} = \frac{f_{tn+t-1}}{f_{t-1}}.$$

For a combinatorial proof of that, see [14].

5.4 Multichoosing

In this section, we examine identities involving the quantity $\left(\binom{n}{k}\right)$, spoken "$n$ multichoose k", which counts the ways to select k objects from a set of n elements, where order is not important, but repetition is allowed. The 20 possible *multisubsets* of size 3 that can be created from $\{1, 2, 3, 4\}$ are illustrated in Figure 5.2. By contrast $\binom{n}{k}$ counts the same

$\left(\!\binom{4}{3}\!\right) = 20$				$\binom{4}{3} = 4$
$\{1,1,1\}$	$\{1,2,3\}$	$\{2,2,2\}$	$\{2,4,4\}$	$\{1,2,3\}$
$\{1,1,2\}$	$\{1,2,4\}$	$\{2,2,3\}$	$\{3,3,3\}$	$\{1,2,4\}$
$\{1,1,3\}$	$\{1,3,3\}$	$\{2,2,4\}$	$\{3,3,4\}$	$\{1,3,4\}$
$\{1,1,4\}$	$\{1,3,4\}$	$\{2,3,3\}$	$\{3,4,4\}$	$\{2,3,4\}$
$\{1,2,2\}$	$\{1,4,4\}$	$\{2,3,4\}$	$\{4,4,4\}$	

Figure 5.2. 3-multisubsets and 3-subsets of $\{1,2,3,4\}$.

situation where repetition is not allowed. The four possible 3-subsets of $\{1,2,3,4\}$ are given in Figure 5.2. Alternately, $\left(\!\binom{n}{k}\!\right)$ counts the nonnegative integer solutions to

$$x_1 + x_2 + \cdots + x_n = k.$$

For $1 \leq i \leq n$, x_i counts the number of times the ith object is chosen.

Here are some of the ways that we like to interpret $\left(\!\binom{n}{k}\!\right)$.

Elections: $\left(\!\binom{n}{k}\!\right)$ counts the ways that k votes can be allocated to n candidates. Here x_i counts the number of votes received by candidate i.

Buckets of ice cream: $\left(\!\binom{n}{k}\!\right)$ counts the ways to choose k scoops of ice cream from n possible flavors, where repetition of flavors is allowed, and the order of the scoops in the bucket is not important. Here x_i denotes how many times flavor i is chosen.

Nondecreasing sequences: $\left(\!\binom{n}{k}\!\right)$ counts the positive integer sequences $a_1, a_2, \ldots, a_k$ where $1 \leq a_1 \leq a_2 \leq \cdots \leq a_k \leq n$. Here x_i denotes the number of i's in the sequence.

Nerds and kandies: $\left(\!\binom{n}{k}\!\right)$ counts ways that we can allocate k *identical* pieces of kandy to n hungry nerds. Nerds may receive any number of kandies, including possibly zero. Here x_i denotes the number of kandies given to nerd i. (We apologize for the intentional misspelling, but it does help you remember which is n and which is k!)

Conveniently, $\left(\!\binom{n}{k}\!\right)$ can be expressed in terms of binomial coefficients. We present three different proofs of the fundamental identity:

Identity 143 *For $k, n \geq 0$ and $k \geq 0$,*

$$\left(\!\binom{n}{k}\!\right) = \binom{n+k-1}{k}.$$

Question: How many ways can we allocate k identical kandies to n nerds?

Answer 1: By definition, $\left(\!\binom{n}{k}\!\right)$.

Answer 2: We represent each allocation with "stars and bars". Specifically, each allocation can be thought of as an arrangement of k stars (each representing a kandy) and $n-1$ bars which act as dividers between nerds. For example, when allocating ten kandies to four nerds, the arrangement of ten stars and three bars given in Figure 5.3 represents the situation where the nerds 1, 2, 3 and 4 receive $3, 2, 0$ and 5 kandies, respectively. Each such arrangement involves placing $n+k-1$ objects in a row and deciding which k of them will be kandies. (In our example, the kandies were placed in spots $1, 2, 3, 5, 6, 9, 10, 11, 12, 13$.)

$$\frac{\star \quad \star \quad \star \quad | \quad \star \quad \star \quad | \quad | \quad \star \quad \star \quad \star \quad \star \quad \star}{1 \quad 2 \quad 3 \quad 4 \quad 5 \quad 6 \quad 7 \quad 8 \quad 9 \quad 10 \ 11 \ 12 \ 13}$$

Figure 5.3. Multichoosing can be represented using stars and bars.

The same identity can be proved by a one-to-one correspondence.

Set 1: Let S denote the set of integer sequences $a_1, a_2, \ldots, a_k$ where $1 \le a_1 \le a_2 \le \cdots \le a_k \le n$. By our earlier interpretation, $|S| = \left(\!\binom{n}{k}\!\right)$.

Set 2: Let T denote the set of integer sequences $b_1, b_2, \ldots, b_k$ where $1 \le b_1 < b_2 < \cdots < b_k \le n + k - 1$. Each element of T can be thought of as a k-element subset of $\{1, \ldots, n + k - 1\}$, $|T| = \binom{n+k-1}{k}$.

Correspondence: For sequence $(a_1, a_2, \ldots, a_k)$ in S, let $b_i = a_i + i - 1$, for $i = 1 \ldots, k$. It is easy to see that the resulting sequence $(b_1, b_2, \ldots, b_k)$ is in T. For example, when $n = 10$ and $k = 6$, the nondecreasing sequence $1, 1, 2, 3, 5, 8$ is mapped to the increasing sequence $1, 2, 4, 6, 9, 13$. Since this procedure is reversible $(a_i = b_i - i + 1)$, it follows that $|S| = |T|$.

Yet another way to derive Identity 143 is to first prove the following.

Identity 144 *For* $0 \le n \le m$,

$$\left(\!\binom{n}{m-n}\!\right) = \binom{m-1}{n-1}.$$

Question: How many ways can we allocate m votes to n candidates, where every candidate gets at least one vote?

Answer 1: First we give each candidate a single vote (since the votes are identical, there is only one way to do this) then we allocate the remaining $m - n$ votes however we want. Hence there are $\left(\!\binom{n}{m-n}\!\right)$ ways to cast votes for the candidates. Note that this quantity is nonzero only when $n \le m$.

Answer 2: Here we do stars and bars a little differently. We begin with m stars each representing a vote, but since no candidate may leave empty-handed, we can not put two bars in a row, nor can we have any bars at the beginning or end. In other words, we have exactly $m - 1$ places where we can place our $n - 1$ dividers. This can be accomplished $\binom{m-1}{n-1}$ ways. In the example in Figure 5.4, candidates $1, 2, 3$ and 4 get $5, 2, 1$ and 2 votes respectively.

$$\frac{\star \quad \star \quad \star \quad \star \quad \star \quad | \quad \star \quad \star \quad | \quad \star \quad | \quad \star \quad \star}{1 \quad 2 \quad 3 \quad 4 \quad 5 \quad 6 \quad 7 \quad 8 \quad 9 \quad 10 \ 11 \ 12 \ 13}$$

Figure 5.4. Another star bar representation.

By letting $m = n + k$, the last identity simplifies to Identity 143.

Identity 145 *For* $n \ge 1$, $k \ge 0$,

$$\left(\!\binom{n}{k}\!\right) = \left(\!\binom{k+1}{n-1}\!\right).$$

Set 1: Let S denote the set of ways to arrange k stars and $n - 1$ bars. By our earlier interpretation, $|S| = \left(\binom{n}{k}\right)$.

Set 2: Let T denote the set of ways to arrange k bars and $n - 1$ stars. By the same interpretation, $|T| = \left(\binom{k+1}{n-1}\right)$.

Correspondence: By turning stars into bars and bars into stars, we have a one-to-one correspondence between S and T. Hence $\left(\binom{n}{k}\right) = \left(\binom{k+1}{n-1}\right)$.

Not surprisingly, there are many multichoose identities that resemble earlier binomial identities. We begin with the Pascal-like identity:

Identity 146 *For $n \geq 0$, $k \geq 0$ (except $n = k = 0$),*

$$\left(\binom{n}{k}\right) = \left(\binom{n}{k-1}\right) + \left(\binom{n-1}{k}\right).$$

Question: In how many ways can we choose k scoops of ice cream from n different flavors?

Answer 1: By definition, $\left(\binom{n}{k}\right)$.

Answer 2: Condition on whether or not the nth flavor is selected. If so, place one scoop of the nth flavor in the bucket (in one way) and select the remaining $k - 1$ scoops in $\left(\binom{n}{k-1}\right)$ ways. Otherwise, the k scoops of ice cream may be selected from the $n - 1$ other flavors in $\left(\binom{n-1}{k}\right)$ ways.

We encourage the reader to provide similar combinatorial proofs using the other combinatorial interpretations of $\left(\binom{n}{k}\right)$. The next identity may seem wrong at first glance:

Identity 147

$$k\left(\binom{n}{k}\right) = n\left(\binom{n+1}{k-1}\right).$$

Question: How many ways can we create a nondecreasing sequence $1 \leq a_1 \leq a_2 \leq \cdots \leq a_k \leq n$ and underline one of the elements?

Answer 1: By definition, there are $\left(\binom{n}{k}\right)$ ways to create the sequence, then k ways to choose the underlined term. Hence there are $k\left(\binom{n}{k}\right)$ such sequences.

Answer 2: First determine the value that will be underlined. There are n choices for this. Suppose that the underlined value is r. Next create a nondecreasing sequence of $k - 1$ elements between 1 and $n + 1$. There are $\left(\binom{n+1}{k-1}\right)$ such sequences. Any rs that are chosen here will go to the left of the underlined r. Any $n + 1$s that are chosen will be converted to rs and repositioned to the right of the underlined r. Hence there are $n\left(\binom{n+1}{k-1}\right)$ such sequences altogether. For example, if $n = 5$, $k = 9$, and our underlined value is $r = 2$, then the 8-sequence $1, 1, 2, 3, 3, 5, 6, 6$ generates the 9-sequence $1, 1, 2, \underline{2}, 2, 2, 3, 3, 5$.

Identity 148 *For $k \geq 1$,*

$$\left(\binom{n}{k}\right) = \sum_{m=1}^{n} \left(\binom{m}{k-1}\right).$$

Question: How many ways can we create a nondecreasing sequence $1 \leq a_1 \leq a_2 \leq \cdots \leq a_k \leq n$?

Answer 1: As usual, $\left(\binom{n}{k}\right)$.

Answer 2: Condition on a_k, the largest element of the sequence. For $1 \leq m \leq n$, the number of such k-sequences where $a_k = m$ equals the number of $(k-1)$-sequences $1 \leq a_1 \leq a_2 \leq \cdots \leq a_{k-1} \leq m$. Since there are, by definition, $\left(\binom{m}{k-1}\right)$ such sequences, then there are $\sum_{m=1}^{n} \left(\binom{m}{k-1}\right)$ such sequences altogether.

Although there is no closed form for $\sum_{k=0}^{m} \binom{n}{k}$, we do have

Identity 149 *For $n \geq 0$,*

$$\sum_{k=0}^{m} \left(\binom{n}{k}\right) = \left(\binom{n+1}{m}\right).$$

Question: In how many ways can m votes be allocated to $n+1$ candidates?

Answer 1: As usual, $\left(\binom{n+1}{m}\right)$.

Answer 2: Condition on the number of votes given to candidates 1 through n. If the first n candidates receive a total of k votes for $0 \leq k \leq m$, then there are $\left(\binom{n}{k}\right)$ ways to allocate these votes, and candidate $n+1$ receives the remaining $m-k$ votes. Altogether there are $\sum_{k=0}^{m} \left(\binom{n}{k}\right)$ allocations.

Our final identity in this section states that

$$\left(\binom{n}{k}\right) = \sum_{m=0}^{n} \binom{n}{m}\binom{k-1}{m-1},$$

or equivalently,

Identity 150 *For $n \geq 0$,*

$$\left(\binom{n}{k}\right) = \sum_{m=0}^{n} \binom{n}{m}\left(\binom{m}{k-m}\right).$$

Question: In how many ways can k identical kandies be allocated to n nerds?

Answer 1: As usual, $\left(\binom{n}{k}\right)$.

Answer 2: Condition on how many nerds receive any kandies at all. For $0 \leq m \leq n$, if there are to be exactly m nerds who receive kandies, then there are $\binom{n}{m}$ ways to select them, then one way to give each a kandy, then $\left(\binom{m}{k-m}\right)$ ways to allocate the remaining $k-m$ kandies to them.

If you are still "hungry" for more multichoose identities, then feast on some of our exercises!

5.5   Odd Numbers in Pascal's Triangle

We conclude this chapter with a pretty combinatorial proof concerning binomial coefficients. If we examine Pascal's Triangle, it appears that the number of odd integers in each row is always a power of 2. More precisely we prove the following amazing theorem.

Theorem 8 *For $n \geq 0$, the number of odd integers in the nth row of Pascal's triangle is equal to 2^b where b is the number of 1s in the binary expansion of n.*

For example, since $76 = 64 + 8 + 4 = (1001100)_{\text{base 2}}$, then row 76 of Pascal's triangle has $2^3 = 8$ odd numbers. In other words, there are eight values of k for which $\binom{76}{k}$ is odd.

A more general result, due to Lucas, is presented in the last chapter of this book. To prove this theorem we shall devise a method to determine the parity of $\binom{n}{k}$ for $0 \leq k \leq n$ and then count the ones that are odd.

The proof of Theorem 8 makes frequent use of a simple numerical fact easily proved by the examining the equation $a = br$.

Lemma 9 *Let r, a, b be integers where $r = \frac{a}{b}$. If a is even and b is odd, then r is even. If a is odd and b is odd, then r is odd.*

The next lemma provides a fast method for determining the parity of $\binom{n}{k}$.

Lemma 10 *If n is even and k is odd, then $\binom{n}{k}$ is even. Otherwise,*

$$\binom{n}{k} \equiv \binom{\lfloor n/2 \rfloor}{\lfloor k/2 \rfloor} \pmod 2.$$

Consequently, except when n is even and k is odd, $\binom{n}{k}$ has the same parity as $\binom{n/2}{k/2}$ where we round $n/2$ and $k/2$ down to the nearest integer, if necessary. For example, we have

$$\binom{57}{37} \equiv \binom{28}{18} \equiv \binom{14}{9} \pmod 2$$

and $\binom{14}{9}$ is even since 14 is even and 9 is odd. Thus $\binom{57}{37}$ is even. On the other hand,

$$\binom{57}{25} \equiv \binom{28}{12} \equiv \binom{14}{6} \equiv \binom{7}{3} \equiv \binom{3}{1} \equiv \binom{1}{0} \equiv 1 \pmod 2.$$

Hence, $\binom{57}{25}$ is odd.

Proof of Lemma 10.

   Case 1: n is even and k is odd. From Identity 130, the integer

$$\binom{n}{k} = \frac{n\binom{n-1}{k-1}}{k}$$

has an even numerator and an odd denominator. Hence by Lemma 9, $\binom{n}{k}$ must be even.

   Case 2: n is even and k is even. We perform the non-combinatorial act of expressing $\binom{n}{k}$ in terms of its formula, then separate the odd numbers from the even numbers. For a more combinatorial derivation, see Identity 224. Consequently,

$$\binom{n}{k} = \frac{n(n-1)(n-2)\cdots(n-k+1)}{1\cdot 2\cdot 3\cdots k}$$

$$= \frac{(n-1)(n-3)\cdots(n-k+1)}{1\cdot 3\cdot 5\cdots(k-1)}\cdot\frac{n(n-2)(n-4)\cdots(n-k+2)}{2\cdot 4\cdot 6\cdots k}$$

$$= \frac{(n-1)(n-3)\cdots(n-k+1)}{1\cdot 3\cdot 5\cdots(k-1)}\cdot\frac{2^{\frac{k}{2}}\cdot\frac{n}{2}(\frac{n}{2}-1)(\frac{n}{2}-2)\cdots(\frac{n}{2}-\frac{k}{2}+1)}{2^{\frac{k}{2}}\cdot 1\cdot 2\cdot 3\cdots\frac{k}{2}}$$

$$= \frac{(n-1)(n-3)\cdots(n-k+1)\cdot\binom{n/2}{k/2}}{1\cdot 3\cdot 5\cdots(k-1)}.$$

Now the resulting denominator is certainly odd, as are all but the last term of the numerator. Hence by Lemma 9, the parity of $\binom{n}{k}$ will be the same as the parity of $\binom{n/2}{k/2}$. Thus

$$\binom{n}{k} \equiv \binom{n/2}{k/2} = \binom{\lfloor n/2\rfloor}{\lfloor k/2\rfloor} \quad (\mathrm{mod}\ 2),$$

as desired.

Case 3: n is odd and k is odd. We again use Identity 130 and Lemma 9 to get

$$\binom{n}{k} = \frac{n\binom{n-1}{k-1}}{k} \equiv \binom{n-1}{k-1} \quad (\mathrm{mod}\ 2).$$

But since $n-1$ and $k-1$ are both even, Case 2 implies $\binom{n-1}{k-1} \equiv \binom{(n-1)/2}{(k-1)/2}$ (mod 2). Thus $\binom{n}{k}$ has the same parity as $\binom{\lfloor n/2\rfloor}{\lfloor k/2\rfloor}$ (mod 2), as desired.

Case 4: n is odd and k is even. Identity 151 in the exercises shows that $(n-k)\binom{n}{k} = n\binom{n-1}{k}$. Consequently, arguing as in Case 3, we have

$$\binom{n}{k} = \frac{n\binom{n-1}{k}}{n-k} \equiv \binom{n-1}{k} \equiv \binom{(n-1)/2}{k/2} = \binom{\lfloor n/2\rfloor}{\lfloor k/2\rfloor} \quad (\mathrm{mod}\ 2),$$

as desired.   ◇

So how does this pertain to our original theorem? We need just a few facts about binary representations. Recall that if x has binary representation $(b_t b_{t-1}\cdots b_1 b_0)_2$, where $b_i = 0$ or 1, then $x = b_t 2^t + b_{t-1}2^{t-1} + \cdots + b_1 2^1 + b_0$. Hence the parity of x is determined by b_0 and $\lfloor x/2\rfloor = b_t 2^{t-1} + b_{t-1}2^{t-2} + \cdots + b_1$ has binary representation $(b_t b_{t-1}\cdots b_2 b_1)_2$.

This allows us to easily apply Lemma 10 when n and k are written in binary. For example, let us evaluate the parity of $\binom{76}{52}$, when both numbers are written in binary.

$$76 = 64 + 8 + 4 = (1001100)_2$$

$$52 = 32 + 16 + 4 = (0110100)_2$$

The leading zero for 52's binary expansion is inserted so that both numbers have the same length. Now since both binary representations end in 0, we must have an $\binom{\mathrm{even}}{\mathrm{even}}$ situation. Hence by repeated use of Lemma 10, this has the same parity as

$$\binom{(100110)_2}{(011010)_2} \equiv \binom{(10011)_2}{(01101)_2} \equiv \binom{(1001)_2}{(0110)_2} \equiv \binom{(100)_2}{(011)_2} \quad (\mathrm{mod}\ 2).$$

$$\binom{76}{52} = \binom{(\mathbf{1001}100)_2}{(\mathbf{0110}100)_2} \qquad \binom{76}{12} = \binom{(1001100)_2}{(0001100)_2}$$

Figure 5.5. $\binom{76}{52}$ is even, but $\binom{76}{12}$ is odd.

Since the last quantity is $\binom{\text{even}}{\text{odd}}$, it follows from Lemma 10 that $\binom{76}{52}$ must be even. In general, $\binom{n}{k}$ will be even if and only if we eventually reduce to $\binom{\text{even}}{\text{odd}}$, which occurs if and only if a 1 appears directly below a 0 in the binary expansions of k and n respectively. See Figure 5.5.

Hence for $\binom{76}{k}$ to be odd,

$$76 = (1\ 0\ 0\ 1\ 1\ 0\ 0)_2$$

k must be of the form

$$k = (x\ 0\ 0\ y\ z\ 0\ 0)_2,$$

where x, y, z can each be either 0 or 1. Thus there are $2^3 = 8$ values of k for which $\binom{76}{k}$ will be odd. By the same reasoning, the number of ways for $\binom{n}{k}$ to be odd is 2^j, where j is the number of 1s in the binary expansion of n.

In fact, the proof tells you exactly which values of k produce an odd number. For 76, they are:

$$64 + 8 + 4 = 76$$

$$64 + 8\ \ \ \ = 72$$

$$64\ \ \ \ + 4 = 68$$

$$64\ \ \ \ \ \ \ \ = 64$$

$$8 + 4 = 12$$

$$8\ \ \ \ = 8$$

$$4 = 4$$

$$0 = 0$$

A generalization of Theorem 8 modulo any prime and two different combinatorial proofs, appears in the section of Chapter 8 called Lucas' Theorem.

5.6 Notes

For innumerably more binomial coefficient identities, proved many different ways, see Riordan's *Combinatorial Identities* [46], Graham, Knuth, and Patashnik's *Concrete Mathematics* [28], or Wilf's *generatingfunctionology* [61]. The proof of odd numbers in Pascal's Triangle is based on the approach presented in Pólya, Tarjan, and Woods [41].

5.7   Exercises

Prove each of the identities below by a direct combinatorial argument.

Identity 151 *For $n \geq k \geq 0$, $(n-k)\binom{n}{k} = n\binom{n-1}{k}$.*

Identity 152 *For $n \geq 2$, $k(k-1)\binom{n}{k} = n(n-1)\binom{n-2}{k-2}$.*

Identity 153 *For $n \geq 3$, $\sum_{k \geq 0} k(k-1)(k-2)\binom{n}{k} = n(n-1)(n-2)\binom{n-3}{3}$.*

Identity 154 *For $n \geq 4$, $\binom{\binom{n}{2}}{2} = 3\binom{n}{4} + 3\binom{n}{3}$.*

Identity 155 *For $0 \leq m \leq n$, $\sum_{k \geq 0} \binom{n}{k}\binom{k}{m} = \binom{n}{m} 2^{n-m}$.*

Identity 156 *For $0 \leq m < n$, $\sum_{k \geq 0} \binom{n}{2k}\binom{2k}{m} = \binom{n}{m} 2^{n-m-1}$.*

Identity 157 *For $m, n \geq 0$, $\sum_{k \geq 0} \binom{m}{k}\binom{n}{k} = \binom{m+n}{n}$.*

Identity 158 *For $m, n \geq 0$, $\sum_{k \geq 0} \binom{n}{k}\binom{n-k}{m-k} = \binom{n}{m} 2^m$.*

Identity 159 *For $n \geq 1$, $\sum_{k \geq 0} k\binom{n}{k}^2 = n\binom{2n-1}{n-1}$.*

Identity 160 *For $n \geq 0$, $\sum_{k=0}^{n} \binom{n}{k}^2 = \binom{2n}{n}$.*

Identity 161 *For $n \geq 0$, $\sum_{k \geq 0} \binom{n}{2k}\binom{2k}{k} 2^{n-2k} = \binom{2n}{n}$.*

The next identity can be proved using binomial or multinomial coefficient interpretations.

Identity 162 *For $m, n \geq 0$, $\sum_{k=0}^{m} \binom{n+k}{k} = \binom{n+m+1}{m}$.*

Identity 163 *For $t \geq 1, 0 \leq c \leq n$, $(G_1 f_t)^c G_{t+1}^{n-c}$ equals*

$$\sum_{x_1 \geq 0} \sum_{x_2 \geq 0} \cdots \sum_{x_t \geq 0} \binom{n-c}{x_1}\binom{n-x_1}{x_2}\binom{n-x_2}{x_3} \cdots \binom{n-x_{t-1}}{x_t} G_0^{x_1} G_1^{n-x_1}$$

where G_j is the jth element of the Gibonacci sequence beginning with G_0 and G_1.

Identity 164 *For $n, k \geq 0$, $\left(\!\binom{n}{2k+1}\!\right) = \sum_{m=1}^{n} \left(\!\binom{m}{k}\!\right)\left(\!\binom{n-m+1}{k}\!\right)$.*

Identity 165 *For $n \geq 0$, $\sum_{k=0}^{n} \binom{n+k}{2k} = f_{2n}$.*

Identity 166 *For $n \geq 1$, $\sum_{k=0}^{n-1} \binom{n+k}{2k+1} = f_{2n-1}$.*

Other Exercises

1. Prove for $n \geq 0, m \geq 1$, that $\sum_{k \geq 0} k \binom{n}{k} \binom{m}{m-k} = n \binom{n+m-1}{m-1}$. Then apply the same logic to arrive at a closed form for $\sum_{k \geq 0} \binom{k}{r} \binom{n}{k} \binom{m}{m-k}$.

2. Many combinatorial proofs for binomial coefficients can also be done by *path counting*. Prove that the number of ways to walk from the point $(0,0)$ to the point (a,b) such that every step is one unit to the right or one unit up is $\binom{a+b}{a}$.

3. Combinatorially prove the identities below by path counting arguments.

   (a) For $a, b > 0$, $\binom{a+b}{a} = \binom{a+b-1}{a} + \binom{a+b-1}{a-1}$.

   (b) For $a, b \geq 0$, $\sum_{k=0}^{a} \binom{a}{k} \binom{b}{a-k} = \binom{a+b}{a}$.

   (c) For $0 \leq s \leq a$, $\sum_{k=0}^{s} \binom{s}{k} \binom{a+b-s}{a-k} = \binom{a+b}{a}$.

   (d) For $a, b \geq 0$, $\sum_{k=0}^{b} \binom{a+k}{a} = \binom{a+b+1}{a+1}$.

   (e) For $0 \leq s \leq a$ and $b \geq 0$, $\sum_{m=s}^{s+b} \binom{m}{s} \binom{a+b-m}{a-s} = \binom{a+b+1}{a}$.

   (f) This last identity only looks simple. For $n \geq 0$, $\sum_{k=0}^{n} \binom{2k}{k} \binom{2n-2k}{n-k} = 4^n$.

4. Catalan numbers. Prove that the number of paths from $(0,0)$ to (n,n) that never go above the main diagonal $y = x$ is $\frac{1}{n+1} \binom{2n}{n}$.

5. Partitions of integers. Let $\pi(n)$ count the ways that the integer n can be expressed as the sum of positive integers, written in non-increasing order. Thus $\pi(4) = 5$, since 4 can be expressed as $4 = 3 + 1 = 2 + 2 = 2 + 1 + 1 = 1 + 1 + 1 + 1$. Prove that the number of integer partitions with at most a positive parts, all of which are at most b, is $\binom{a+b}{a}$. (Example: When $a = 2, b = 3$, the ten partitions are: $3 + 3, 3 + 2, 3 + 1, 3, 2 + 2, 2 + 1, 2, 1 + 1, 1, \phi$.)

6. An ordered partition or (composition) of n does not require the summands to be in non-increasing order. For instance 4 has eight ordered partitions: $4 = 3 + 1 = 1 + 3 = 2 + 2 = 2 + 1 + 1 = 1 + 2 + 1 = 1 + 1 + 2 = 1 + 1 + 1 + 1$. Prove that the number of ordered partitions of n with exactly k parts is $\binom{n-1}{k-1}$ and the total number of ordered partitions of n is 2^{n-1}.

CHAPTER 6

Alternating Sign Binomial Identities

6.1 Parity Arguments and Inclusion-Exclusion

In the last chapter, we proved, for $n > 0$, $\sum_{k \geq 0} \binom{n}{2k} = 2^{n-1}$. Since $\sum_{k \geq 0} \binom{n}{k} = 2^n$, this implies that half of all subsets of $\{1, \ldots, n\}$ are even. Consequently

$$\sum_{k \geq 0} \binom{n}{2k} = \sum_{k \geq 0} \binom{n}{2k+1}.$$

This suggests that there should be a simple one-to-one correspondence between the even subsets of $\{1, 2, \ldots, n\}$ and the odd ones. We begin with a bijective proof of this fact.

Identity 167 *For $n > 0$,*

$$\sum_{k=0}^{n} \binom{n}{k} (-1)^k = 0.$$

Set 1: Let $\mathcal{E}$ denote the set of even subsets $\{a_1, \ldots, a_k\}$ of $\{1, \ldots, n\}$ where k is an even number. This set has size $\sum_{k \text{ even}} \binom{n}{k}$.

Set 2: Let $\mathcal{O}$ denote the set of odd subsets $\{a_1, \ldots, a_k\}$ of $\{1, \ldots, n\}$ where k is an odd number. This set has size $\sum_{k \text{ odd}} \binom{n}{k}$.

Correspondence: For $X = \{a_1, a_2, \ldots, a_k\} \in \mathcal{E}$, where $1 \leq a_1 < a_2 < \cdots < a_k \leq n$, consider the *symmetric difference* Y of X with $\{n\}$. Then

$$Y = X \oplus \{n\}$$
$$= \begin{cases} \{a_1, a_2, \ldots, a_k, n\} & \text{if } n \notin X, \\ \{a_1, a_2, \ldots, a_{k-1}\} & \text{if } n \in X. \end{cases}$$

In other words, if n is not in X, then $X \oplus \{n\}$ puts n into X, and if n is in X, then $X \oplus \{n\}$ removes it from X. Notice that $X \oplus \{n\}$ has either one more or one fewer elements than X, so it must have an odd number of elements. Further notice that $(X \oplus \{n\}) \oplus \{n\} = X$, so the correspondence is reversible. Thus $|\mathcal{E}| = |\mathcal{O}|$, as desired.

If we sum only some of these numbers, we have the more general

Identity 168 *For $m \geq 0$ and $n > 0$,*

$$\sum_{k=0}^{m} \binom{n}{k} (-1)^k = (-1)^m \binom{n-1}{m}.$$

Notice that when $m \geq n$, this reduces to the previous identity, so we shall only consider the case where $0 \leq m < n$. The proof is practically the same as before, but now our correspondence is *almost* one-to-one.

Set 1: Let $\mathcal{E}$ denote the set of even subsets $\{a_1, \ldots, a_k\}$ of $\{1, \ldots, n\}$ where k is an arbitrary even number that is less than or equal to m. This set has size $\sum_{k \text{ even} \geq 0}^{m} \binom{n}{k}$.

Set 2: Let $\mathcal{O}$ denote the set of odd subsets $\{a_1, \ldots, a_k\}$ of $\{1, \ldots, n\}$ where k is an arbitrary odd number that is less than or equal to m. This set has size $\sum_{k \text{ odd} \geq 1}^{m} \binom{n}{k}$.

Correspondence: As in the last proof, the symmetric difference $X \oplus \{n\}$ will do the trick. We separately prove the cases where m is even and m is odd.

When m is even and X is a subset in $\mathcal{E}$, $X \oplus \{n\}$ belongs to $\mathcal{O}$, provided that $X \oplus \{n\}$ does not have more than m elements. The only unmatched subsets occur in the $\binom{n-1}{m}$ instances where $|X| = m$ and $n \notin X$, since $X \oplus \{n\}$ contains $m + 1$ elements. Thus when m is even, $|\mathcal{E}| = |\mathcal{O}| + \binom{n-1}{m}$.

When m is odd and X is a subset in $\mathcal{E}$, $X \oplus \{n\}$ is always defined, but we miss those members of $\mathcal{O}$ that contain m elements from $\{1, \ldots, n-1\}$ since such a set can only be hit by an $m + 1$ element subset of $\{1, \ldots, n\}$ that contains n. Here, $|\mathcal{O}| = |\mathcal{E}| + \binom{n-1}{m}$.

Combining the even and odd cases, we have that $|\mathcal{E}| - |\mathcal{O}| = (-1)^m \binom{n-1}{m}$, as desired.

Identity 167 can be used to prove the useful **principle of inclusion-exclusion.**

Theorem 11 *For finite sets $A_1, A_2, \ldots, A_n$, $|A_1 \cup A_2 \cup \cdots \cup A_n|$ is equal to*

$$\sum_{1 \leq i \leq n} |A_i| \ - \sum_{1 \leq i < j \leq n} |A_i \cap A_j|$$

$$+ \sum_{1 \leq i < j < k \leq n} |A_i \cap A_j \cap A_k| \ - \cdots + (-1)^n |A_1 \cap A_2 \cap \cdots \cap A_n|.$$

When $n = 2$, the principle of inclusion-exclusion says that

$$|A_1 \cup A_2| = |A_1| + |A_2| - |A_1 \cap A_2|,$$

easily seen by the Venn Diagram in Figure 6.1 since $|A_1| + |A_2|$ counts all of the points in $A_1 \cup A_2$, but the points in $A_1 \cap A_2$ are counted twice. Hence these terms get *uncounted* once by the $-|A_1 \cap A_2|$ term.

Likewise, when $n = 3$, inclusion-exclusion says $|A_1 \cup A_2 \cup A_3|$ equals

$$|A_1| + |A_2| + |A_3| \ - (|A_1 \cap A_2| + |A_1 \cap A_3| + |A_2 \cap A_3|) + |A_1 \cap A_2 \cap A_3|.$$

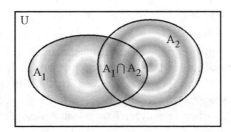

Figure 6.1. With inclusion-exclusion, every element in the union is counted once.

The elements that are in exactly one set are counted by the sum $|A_1| + |A_2| + |A_3|$ but elements in more than one set are overcounted. To remedy this, we subtract the next three terms $(-|A_1 \cap A_2| - |A_1 \cap A_3| - |A_2 \cap A_3|)$. Now each point in exactly two sets has been counted once [more precisely, they have been counted two times and uncounted one time], but points in all three sets have been counted zero times [counted three times then uncounted three times] so the last term $|A_1 \cap A_2 \cap A_3|$ is added to remedy this. Consequently, every element that is in at least one set is counted exactly once (and elements in no sets are never counted.)

To see that inclusion-exclusion works for any n, consider an element x in exactly k of the n sets for some $1 \leq k \leq n$. Then $|A_1| + |A_2| + \cdots + |A_n|$ will count that element exactly k times. Now x appears in exactly $\binom{k}{2}$ pairs of sets, so $\sum_{1 \leq i < j \leq n} |A_i \cap A_j|$ will uncount it $\binom{k}{2}$ times. Then $\sum_{1 \leq i < j < k \leq n} |A_i \cap A_j \cap A_k|$ will count it $\binom{k}{3}$ times, and so on. Overall, element x will be counted

$$\binom{k}{1} - \binom{k}{2} + \binom{k}{3} - \binom{k}{4} + \cdots + (-1)^{k+1}\binom{k}{k}$$

times. By Identity 167, this quantity equals $\binom{k}{0} = 1$. Hence every element in $|A_1 \cup A_2 \cup \cdots \cup A_n|$ is counted exactly once.

We have just used Identity 167 to prove the principle of inclusion-exclusion. And yet, satisfyingly, we can also use inclusion-exclusion to prove Identity 167.

Question: Suppose $A_1 = A_2 = \cdots = A_n = \{1\}$. Find $|A_1 \cup A_2 \cup \cdots \cup A_n|$.

Answer 1: Clearly, $A_1 \cup A_2 \cup \cdots \cup A_n = \{1\}$ has size 1.

Answer 2: Using inclusion-exclusion, for $1 \leq k \leq n$, each intersection of k subsets contributes 1 to our alternating sum. Thus our union has size

$$\binom{n}{1} - \binom{n}{2} + \binom{n}{3} - \cdots + (-1)^n \binom{n}{n}.$$

We can learn even more about inclusion-exclusion by Identity 168. First observe that if $A_1, \ldots, A_n$ are disjoint sets, then $|A_1 \cup \cdots \cup A_n| = |A_1| + \cdots + |A_n|$. But if any element is in at least two sets, then clearly $\sum_{i=1}^{n} |A_i|$ *overcounts* $|A_1 \cup \cdots \cup A_n|$. More generally, suppose that we apply the first m steps of inclusion-exclusion to evaluate $|A_1 \cup \cdots \cup A_n|$. By the proof of inclusion-exclusion, any element that appears in exactly k sets will be counted

$$\binom{k}{1} - \binom{k}{2} + \binom{k}{3} - \binom{k}{4} + \cdots + (-1)^{m+1}\binom{k}{m}$$

times. But by Identity 168, this quantity simplifies to $\binom{k}{0} + (-1)^{m+1}\binom{k-1}{m} = 1 + (-1)^{m+1}\binom{k-1}{m}$. Note that when $k \leq m$, then $\binom{k-1}{m} = 0$, and the element is counted exactly once, as desired. However, if $k > m$, then that element will either be overcounted or undercounted $\binom{k-1}{m}$ times, depending on the parity of m. Consequently, if any element appears in more than m of the sets $A_1, \ldots, A_n$, then the first m steps of inclusion-exclusion will necessarily *overcount* $|A_1 \cup \cdots \cup A_n|$ when m is odd, and will necessarily *undercount* $|A_1 \cup \cdots \cup A_n|$ when m is even. This is sometimes referred to as *Bonferroni's Inequality*.

6.2 Alternating Binomial Coefficient Identities

In this section, we shall prove some interesting binomial coefficient identities with alternating terms, either by demonstrating almost one-to-one correspondences between two sets or by invoking the principle of inclusion-exclusion. Some of these identities utilize the *Kronecker delta* function $\delta_{n,m}$ which equals 1 whenever $n = m$, and is 0 otherwise. We begin with the following generalization of Identity 167.

Identity 169 *For $m, n \geq 0$,*

$$\sum_{k=0}^{n} \binom{n}{k}\binom{k}{m}(-1)^k = (-1)^n \delta_{n,m}.$$

Set 1: For $n, m \geq 0$, let $\mathcal{E}$ denote the set of ordered pairs (S, T) where $T \subseteq S \subseteq \{1, \ldots, n\}$, where $|T| = m$ and $|S|$ is even. For even numbers k, $\mathcal{E}$ contains $\binom{n}{k}\binom{k}{m}$ elements where $|S| = k$, and therefore contains $\sum_k$ even $\binom{n}{k}\binom{k}{m}$ elements altogether.

Set 2: For $n, m \geq 0$, let $\mathcal{O}$ denote the set of ordered pairs (S, T) where $T \subseteq S \subseteq \{1, \ldots, n\}$, where $|T| = m$ and $|S|$ is odd. For odd numbers k, $\mathcal{O}$ contains $\binom{n}{k}\binom{k}{m}$ elements where $|S| = k$, and therefore contains $\sum_k$ odd $\binom{n}{k}\binom{k}{m}$ elements altogether.

Correspondence: First note that when $n < m$, both $\mathcal{E}$ and $\mathcal{O}$ are empty, so the identity is trivially true. When $n = m$ is even, then $\mathcal{E}$ contains one element, namely $T = S = \{1, \ldots, n\}$, and $\mathcal{O}$ is empty. Likewise, when $n = m$ is odd, then $\mathcal{E}$ is empty and $\mathcal{O}$ contains one element. Either way, the sizes of $\mathcal{E}$ and $\mathcal{O}$ differ by exactly one element, as predicted by the identity.

When $n > m$, we establish a one-to-one correspondence between $\mathcal{E}$ and $\mathcal{O}$ as follows. For any (S, T) pair, let x be the largest element of $\{1, \ldots, n\}$ that is not in T. (Such an x must exist since $n > m$. Now if $x \in S$, we remove it. If $x \notin S$, we put it in. In other words, we associate (S, T) with $(S \oplus x, T)$. Since $|S|$ and $|S \oplus x|$ have opposite parity, we have a one-to-one correspondence between $\mathcal{E}$ and $\mathcal{O}$.

The same idea can be applied to the next identity where we first choose a subset $S \subseteq \{1, \ldots, n\}$ followed by a size m multisubset of S.

Identity 170 *For $n \geq m$,*

$$\sum_{k=0}^{n} \binom{n}{k}\left(\binom{k}{m}\right)(-1)^k = (-1)^n \delta_{n,m}.$$

Set 1: For $n \geq m \geq 0$, let $\mathcal{E}$ denote the set of ways to choose an even number of candidates from $\{1, \ldots, n\}$, then allow m votes to be distributed among them. For an even number of candidates k, $\mathcal{E}$ contains $\binom{n}{k}\left(\binom{k}{m}\right)$ elements, and therefore contains $\sum_{k \text{ even}} \binom{n}{k}\left(\binom{k}{m}\right)$ elements altogether.

Set 2: For $n \geq m \geq 0$, let $\mathcal{O}$ denote the set of ways to choose an odd number of candidates from $\{1, \ldots, n\}$, then allow m votes to be distributed among them. For an odd number of candidates k, $\mathcal{O}$ contains $\binom{n}{k}\left(\binom{k}{m}\right)$ elements, and therefore contains $\sum_{k \text{ odd}} \binom{n}{k}\left(\binom{k}{m}\right)$ elements altogether.

Correspondence: If $m = n$, the situation where all n candidates are chosen and each receives one vote can only occur one way. This is the unmatched situation and it contributes $(-1)^n$ to the summation. Otherwise, we shall pair together elections with even and odd numbers of candidates. For a given element of $\mathcal{E}$, let x denote the largest member of $\{1, \ldots, n\}$ who receives no votes. Since $m \leq n$, such an x exists. If x is not among the candidates, put x on the ballot (although he receives no votes). If x is among the candidates, we may remove candidate x from the ballot and not disrupt any of the votes since x receives no votes. In other words, we are associating the ballot S and its multisubset of votes T with $(S \oplus x, T)$, and the identity follows.

In the next identity, we extend the range of Identity 170 by allowing m to exceed n.

Identity 171 *For any* $m, n \geq 0$,

$$\sum_{k=0}^{n} \binom{n}{k} \left(\binom{k}{m}\right) (-1)^k = (-1)^n \left(\binom{n}{m-n}\right).$$

Here, Set 1 and Set 2 are unchanged from the previous proof. When $m \leq n$, this reduces to the previous identity. When $m > n$, we have the following correspondence.

Correspondence: Here, $\left(\binom{n}{m-n}\right)$ counts (S, T) where all elements of $\{1, \ldots, n\}$ are on the ballot, and each receives at least one vote. These pairs remain unmatched. Otherwise, choosing x as the largest numbered element of n receiving no votes and associating (S, T) with $(S \oplus x, T)$ yields the same correspondence between $\mathcal{E}$ and $\mathcal{O}$.

The next two identities are tiling identities in disguise. Recall from Identity 4 in Chapter 1 that the number of square-domino tilings of an n-board with exactly k dominoes is $\binom{n-k}{k}$. Consequently,

$$\sum_{k \geq 0} (-1)^k \binom{n-k}{k}$$

counts the difference between the number of n-tilings with an even number of dominoes versus those with an odd number of dominoes.

Identity 172 *For* $n \geq 0$,

$$\sum_{k \geq 0} (-1)^k \binom{n-k}{k} = \begin{cases} 1 & \text{if } n \equiv 0 \text{ or } 1 \pmod 6, \\ 0 & \text{if } n \equiv 2 \text{ or } 5 \pmod 6, \\ -1 & \text{if } n \equiv 3 \text{ or } 4 \pmod 6. \end{cases}$$

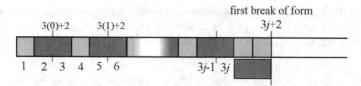

Figure 6.2. When $n \equiv 2 \pmod 3$, an n-board with an even number of dominoes can be easily transformed to an n-board with an odd number of dominoes, and vice versa.

Set 1: Let $\mathcal{E}$ denote the set of n-tilings with an even number of dominoes. Thus $|\mathcal{E}| = \sum_{k \text{ even}} \binom{n-k}{k}$.

Set 2: Let $\mathcal{O}$ denote the set of n-tilings with an odd number of dominoes. Thus $|\mathcal{O}| = \sum_{k \text{ odd}} \binom{n-k}{k}$.

Correspondence: We first find a one-to-one correspondence between $\mathcal{E}$ and $\mathcal{O}$ when $n \equiv 2$ or $5 \pmod 6$, i.e., when $n \equiv 2 \pmod 3$. For a given tiling with an even number of dominoes, find the first breakable cell of the form $3j + 2$. Such a cell must exist since the last cell, cell n, is of this form. To avoid having breaks at cells $3i + 2$ for $i < j$, we must begin with j "square-domino" pairs. See Figure 6.2. Cells $3j + 1$ and $3j + 2$ are either covered by the same domino or by two squares. Our correspondence from $\mathcal{E}$ to $\mathcal{O}$ is as follows: If cells $3j + 1$ and $3j + 2$ are covered by the same domino, then turn that domino into two squares. If cell $3j + 1$ and $3j + 2$ are covered by two squares, then turn those two squares into a domino. Since our tiling remains breakable at cell $3j + 2$ and remains unbreakable at cells $3i + 2$ for $i < j$, it is easily reversed. Consequently, for $n \equiv 2 \pmod 3$, $|\mathcal{E}| = |\mathcal{O}|$.

If $n \not\equiv 2 \pmod 3$, then there is exactly one tiling (belonging to either $\mathcal{E}$ or $\mathcal{O}$, depending on n) that is unbreakable at all cells of the form $3i + 2$. Specifically, if $n = 6q + 1$, then the unmatched tiling consists of $2q$ square-domino pairs followed by a single square. Since the number of dominoes is even, then for $n \equiv 1 \pmod 6$, $|\mathcal{E}| - |\mathcal{O}| = 1$. Likewise, for when $n \equiv 0 \pmod 6$. Similarly, when $n = 6q + 3$ or $n = 6q + 4$, the unmatched tiling, consisting of $2q + 1$ square-domino pairs (followed by a single square when $n = 6q + 4$), belongs to $\mathcal{O}$ since it contains an odd number of dominoes. Here, we have $|\mathcal{E}| - |\mathcal{O}| = -1$.

Next we consider "colored" square-domino tilings of an n-board where we now have two colors for squares (black and white) but just one color for dominoes.

Identity 173 *For $n \geq 0$,*

$$\sum_{k \geq 0} (-1)^k \binom{n-k}{k} 2^{n-2k} = n + 1.$$

Set 1: Let $\mathcal{E}$ denote the set of n-tilings with an even number of dominoes. Once the locations of the k dominoes are determined the remaining $n - 2k$ cells may be covered by black or white squares. Consequently, $|\mathcal{E}| = \sum_{k \text{ even}} \binom{n-k}{k} 2^{n-2k}$.

Set 1: Let $\mathcal{O}$ denote the set of n-tilings with an odd number of dominoes. Here, $|\mathcal{O}| = \sum_{k \text{ odd}} \binom{n-k}{k} 2^{n-2k}$.

Correspondence: Here, we single out the $n + 1$ tilings consisting of k black squares followed by $n - k$ white squares for some $0 \leq k \leq n$. These tilings are in $\mathcal{E}$ since

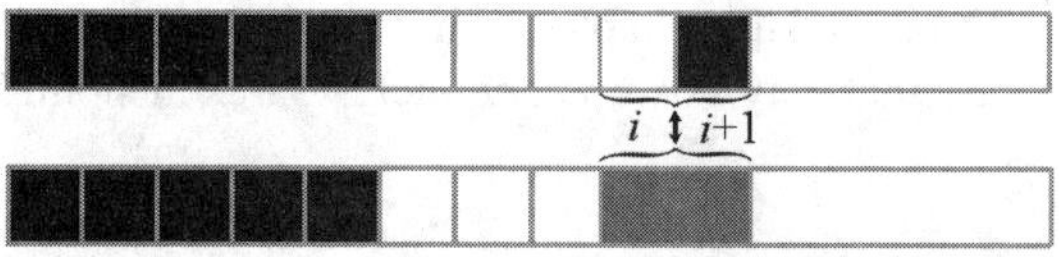

Figure 6.3. The first occurrence of a domino or a white-black pair is converted into the other form.

they contain zero dominoes. All other tilings must contain a domino or a white square followed by a black square (a "white-black" pair). For such a tiling in $\mathcal{E}$, let i denote the first cell for which cells i and $i+1$ are covered by either a domino or by a white-black pair. In the first case, we convert that domino into a white-black pair; in the second case, we convert the white-black pair into a domino. See Figure 6.3. Either way, we change the parity of the number of dominoes. Since the correspondence is easily reversed, we have $|\mathcal{E}| - |\mathcal{O}| = n + 1$.

The previous identity can also be proved using inclusion-exclusion.

Question: How many ways can we tile an n-board using only black and white squares (no dominoes) and with the restriction that no black square is immediately preceded by a white square?

Answer 1: There are $n + 1$ such arrangements, namely those consisting of k black squares followed by $n - k$ white squares for some $0 \le k \le n$.

Answer 2: Since each of the n cells must be covered by a black or white square there are 2^n arrangements, ignoring the restriction. From these we must subtract those arrangements where a white square precedes a black square. For $1 \le i \le n - 1$, let A_i denote the set of all-square arrangements where cell i is white and cell $i + 1$ is black. Consequently, the answer to our question is

$$2^n - |A_1 \cup A_2 \cup \cdots \cup A_{n-1}|.$$

Notice that it is impossible for a tiling to be in both sets A_i and A_{i+1} since cell $i+1$ would have to be both black and white. Consequently, for a tiling to be in k of these sets, it has $2k$ of its cells prescribed (by placing k "white-black" pairs of squares at the prescribed locations) and the remaining $n - 2k$ cells can be covered in 2^{n-2k} ways. See Figure 6.4. The number of ways that k of the sets A_i can be chosen is equal to the number of ways that a traditional uncolored n-board can be tiled with exactly k dominoes, namely $\binom{n-k}{k}$. Consequently, by the principle of inclusion-exclusion, we have that the number of restricted arrangements is $2^n - \sum_{k \ge 1}(-1)^{k-1}\binom{n-k}{k}2^{n-2k}$, or more compactly: $\sum_{k \ge 0}(-1)^k\binom{n-k}{k}2^{n-2k}$.

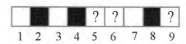

Figure 6.4. A 9-board that belongs to sets A_1, A_3, and A_7 can be covered with black and white squares in 2^3 ways.

Finally, since we INtroduced this chapter with inclusion-exclusion, it seems only fitting that we EXit this chapter that way as well. The inclusion-exclusion formula can be stated more compactly as

$$|A_1 \cup \cdots \cup A_n| = \sum_{s=1}^{n} \sum_{|S|=s} |A_S|(-1)^{s-1},$$

where the second summation is summing over all s-element subsets $S \subseteq \{1, \ldots, n\}$, and A_S consists of those elements that appear in the sets A_i where $i \in S$. (Put another way, $A_S = \cap_{i \in S} A_i$. Note that elements of A_S may also appear in other A_j where $j \notin S$.)

Suppose we are interested in counting the number of elements that appear in at least m of the n sets $A_1, \ldots, A_n$? As it turns out, the inclusion-exclusion formula, with one slight modification, gives us the answer when we only use the sets A_S where $|S| \geq m$. Specifically, for $1 \leq m \leq n$, the number of elements that appear in at least m of the n sets $A_1, \ldots, A_n$ is

$$\sum_{s=m}^{n} \binom{s-1}{m-1} \sum_{|S|=s} |A_S|(-1)^{s-m}.$$

Notice that when $m = 1$, this is the usual inclusion-exclusion formula.

Naturally an element that appears in fewer than m sets is never counted by the above formula. To prove the formula, we need to show that an element in at least m of the above sets is counted exactly once. Since an element that appears in exactly k sets will also appear in $\binom{k}{s}$ sets of size s, our problem reduces to proving that for any $k \geq m$,

$$\sum_{s=m}^{k} \binom{k}{s} \binom{s-1}{m-1} (-1)^{s-m} = 1.$$

Substituting $\binom{k}{s} = \binom{k}{k-s}$, $\binom{s-1}{m-1} = \left(\binom{m}{s-m}\right)$ and letting $d = s - m$ and $y = k - m$, we prove the equivalent identity below:

Identity 174 *For $y, m \geq 0$,*

$$\sum_{d=0}^{y} \binom{m+y}{y-d} \left(\binom{m}{d}\right) (-1)^d = 1.$$

Background: A dessert shop sells m different flavors of ice cream and y different flavors of frozen yogurt. Art and Deena plan to purchase a total of y scoops of ice cream or yogurt with the following restrictions. Art's container will only have distinct flavors. Deena will only select among ice cream flavors (no yogurt) but she is willing to have flavors repeated in her container. The number of ways Art and Deena can leave the dessert shop with Deena's container containing d scoops and Art's container containing $y - d$ scoops is $\binom{m+y}{y-d} \left(\binom{m}{d}\right)$.

Set 1: Let $\mathcal{E}$ denote the ways that Art and Deena can leave the dessert shop where Deena has an even number of scoops. This set has size $\sum_{d \text{ even}} \binom{m+y}{y-d} \left(\binom{m}{d}\right)$.

Set 2: Let $\mathcal{O}$ denote the ways that Art and Deena can leave the dessert shop where Deena has an odd number of scoops. This set has size $\sum_{d \text{ odd}} \binom{m+y}{y-d} \left(\binom{m}{d}\right)$.

Correspondence: We provide an almost one-to-one correspondence between $\mathcal{E}$ and $\mathcal{O}$. The only unmatched allocation in $\mathcal{E}$ occurs when Deena leaves empty-handed and

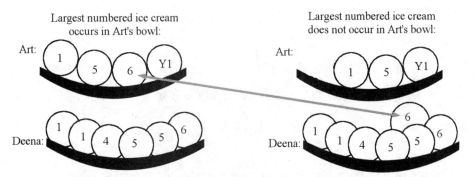

Figure 6.5. Except when all y scoops consist of frozen yogurt in Art's container, we can always transfer the largest numbered ice cream scoop from one container to the other.

Art chose as his y scoops all y yogurt flavors. (This was most inconsiderate on Art's part, since Deena can't even share one of Art's scoops!) Otherwise, at least one of the ice cream flavors (numbered from 1 to m) will appear in at least one container. If i is the largest numbered flavor to appear in at least one container, we change the parity of Deena's container as follows. If flavor i is in Art's container, then he transfers it to Deena's container. (Some might say that Deena is getting her just desserts; others might simply call it just-ice.) Otherwise, Deena has at least one scoop of flavor i and she transfers it to Art's container. Notice that after the transfer, i remains the largest numbered flavor to appear in at least one container, so the process is easily reversed. See Figure 6.5. Hence, except for the one inconsiderate allocation, there are as many even allocations as odd ones.

6.3 Notes

Our approach illustrates the signed involution technique described in Chapter 4 of Stanton and White's *Constructive Combinatorics* [53] which is in turn based on the more general signed set principle of Garsia and Milne [25]. The connection between Garsia and Milne's work and the principle of inclusion-exclusion is explicitly reported by Zeilberger in [63].

6.4 Exercises

1. **Counting Onto Functions** Over a period of m days, in a class of $n > 0$ students, each day a student is selected to lead the class in song. Let $A(m, n)$ count the number of ways that the leaders can be chosen so that every student gets to lead at least once. Use inclusion-exclusion to prove

$$A(m, n) = \sum_{k=0}^{n} \binom{n}{k} (n - k)^m (-1)^k.$$

2. As an immediate corollary of Exercise 1, deduce that when $m < n$,

$$\sum_{k=0}^{n} \binom{n}{k} (n - k)^m (-1)^k = 0.$$

3. As an immediate corollary of Exercise 1, deduce that when $m = n$,

$$\sum_{k=0}^{n} \binom{n}{k} (n-k)^n (-1)^k = n!$$

4. As an immediate corollary of Exercise 1, deduce that when $m = n + 1$,

$$\sum_{k=0}^{n} \binom{n}{k} (n-k)^{n+1} (-1)^k = n\frac{(n+1)!}{2}.$$

5. Now prove Exercise 1 by finding a correspondence between an appropriately chosen even set and odd set.

   Hint: You may wish to first find a bijective proof for Exercise 2 and an almost one-to-one correspondence for Exercise 3.

6. Multiplying Identity 172 by $(-1)^n$ produces the simpler identity below. Prove it by finding the appropriate correspondence.

   Identity 175 *For $n \geq 0$,*

$$\sum_{k \geq 0} (-1)^{n-k} \binom{n-k}{k} = \begin{cases} 1 & \text{if } n \equiv 0 \pmod 3, \\ 0 & \text{if } n \equiv 2 \pmod 3, \\ -1 & \text{if } n \equiv 1 \pmod 3. \end{cases}$$

7. **Bracelet identities** Recall from Chapter 2 that the number of n-bracelets (i.e., circular boards of length n covered with squares and dominoes) with exactly k dominoes is $\frac{n}{n-k} \binom{n-k}{k}$. Using this interpretation, we can derive some interesting alternating binomial coefficient identities. Prove the following identity by finding the appropriate correspondence between even and odd sets.

   Identity 176 *For $n \geq 0$,*

$$\sum_{k \geq 0} (-1)^k \frac{n}{n-k} \binom{n-k}{k} = \begin{cases} 2 & \text{if } n \equiv 0 \pmod 6, \\ 1 & \text{if } n \equiv 1 \text{ or } 5 \pmod 6, \\ -1 & \text{if } n \equiv 2 \text{ or } 4 \pmod 6, \\ -2 & \text{if } n \equiv 3 \pmod 6. \end{cases}$$

   Challenge: Can you find a proof of the above using inclusion-exclusion?

8. Using the principle of inclusion-exclusion, prove

   Identity 177 *For $n \geq 0$,*

$$\sum_{k \geq 0} (-1)^k \frac{n}{n-k} \binom{n-k}{k} 2^{n-2k} = 2.$$

9. Prove the previous identity by finding the appropriate correspondence.

More exercises with alternating sums are given in Chapter 7.

CHAPTER 7

Harmonic and Stirling Number Identities

Definition The *nth harmonic number* is $H_n = 1 + \frac{1}{2} + \frac{1}{3} + \cdots + \frac{1}{n}$. The first few harmonic numbers are $H_1 = 1$, $H_2 = \frac{3}{2}$, $H_3 = \frac{11}{6}$, $H_4 = \frac{25}{12}$.

Definition The *Stirling number of the first kind* $\begin{bmatrix} n \\ k \end{bmatrix}$ counts the number of permutations of n elements with exactly k cycles. Some examples when $k = 2$: $\begin{bmatrix} 1 \\ 2 \end{bmatrix} = 0$, $\begin{bmatrix} 2 \\ 2 \end{bmatrix} = 1$, $\begin{bmatrix} 3 \\ 2 \end{bmatrix} = 3$, $\begin{bmatrix} 4 \\ 2 \end{bmatrix} = 11$, $\begin{bmatrix} 5 \\ 2 \end{bmatrix} = 50$.

Definition The *Stirling number of the second kind* $\begin{Bmatrix} n \\ k \end{Bmatrix}$ counts the number of partitions of $\{1, \ldots, n\}$ into exactly k subsets. Some examples when $k = 2$: $\begin{Bmatrix} 1 \\ 2 \end{Bmatrix} = 0$, $\begin{Bmatrix} 2 \\ 2 \end{Bmatrix} = 1$, $\begin{Bmatrix} 3 \\ 2 \end{Bmatrix} = 3$, $\begin{Bmatrix} 4 \\ 2 \end{Bmatrix} = 7$, $\begin{Bmatrix} 5 \\ 2 \end{Bmatrix} = 15$.

7.1 Harmonic Numbers and Permutations

Harmonic numbers are defined to be partial sums of the harmonic series. That is, for $n \geq 1$, let

$$H_n = 1 + \frac{1}{2} + \frac{1}{3} + \cdots + \frac{1}{n}.$$

The first five harmonic numbers are $H_1 = 1$, $H_2 = 3/2$, $H_3 = 11/6$, $H_4 = 25/12$, $H_5 = 137/60$. For convenience we define $H_0 = 0$. Since the harmonic series diverges, H_n gets arbitrarily large, although it does so quite slowly. For instance, $H_{1,000,000} \approx 14.39$.

Harmonic numbers even appear in real life. If you stack 2-inch long playing cards to overhang the edge of a table as far as possible, the maximum distance that n cards can hang off the edge of the table is H_n [28]. For example, 4 cards can be stacked to extend past the table by just over 2 inches, since $H_4 = \frac{25}{12}$.

It can be shown [28] that H_n is never an integer for $n > 1$. This would suggest that no combinatorial interpretation of them would exist. But after seeing harmonic number identities like the three below, we might think otherwise.

Identity 178 *For $n \geq 1$, $\sum_{k=1}^{n-1} H_k = n H_n - n$.*

Identity 179 *For* $0 \leq m < n$, $\sum_{k=m}^{n-1} \binom{k}{m} H_k = \binom{n}{m+1}(H_n - \frac{1}{m+1})$.

Identity 180 *For* $0 \leq m \leq n$, $\sum_{k=m}^{n-1} \binom{k}{m} \frac{1}{n-k} = \binom{n}{m}(H_n - H_m)$.

Although all of these identities can be proved by algebraic methods, the presence of binomial coefficients suggests that these identities can also be proved combinatorially. Indeed, H_n can certainly be written as a (typically nonreduced) fraction of the form $\frac{a_n}{n!}$. Since the denominator has obvious combinatorial interpretation, it makes sense that the numerator should have one as well. So what are harmonic numbers counting? That will be revealed in the next two sections. First, however, we need to say a bit more about factorials and permutations.

There are many ways to think about $n! = n(n-1)(n-2) \cdots 1$. For instance, as we saw in the proof of Identity 125, $n!$ counts the number of ways that 1 through n can be arranged since there are n choices for the first number, $n-1$ choices for the second number and so on. Such an arrangement is called a *permutation* of $1, 2, \ldots, n$. Using this interpretation, we can easily explain

Identity 181 *For* $n \geq 1$, $\sum_{k=1}^{n-1} k \cdot k! = n! - 1$.

Question: How many ways can the numbers 1 through n be arranged, where we exclude the natural permutation 1 2 3 ... n?

Answer 1: There are $n! - 1$ such permutations.

Answer 2: Condition on the first number that is not in its natural position. For $1 \leq k \leq n-1$, how many permutations have $n-k$ as the first number to differ from its natural position? Such a permutation begins 1 2 3 ... $n-k-1$ followed by one of k numbers from the set $\{n-k+1, n-k+2, \ldots, n\}$. The remaining k numbers (now including the number $n-k$) can be arranged $k!$ ways. Thus there are $k \cdot k!$ ways for $n-k$ to be the first misplaced number. Summing over all feasible values of k yields the left side of the identity.

Here is another way to look at permutations that will be more useful to us in this chapter. The arrangement 265431 can be described as follows: The number 1 is sent to location 6; number 6 is sent to location 2; number 2 is sent to location 1. Numbers 3 and 5 have swapped positions, and number 4 has stayed in its natural position. We could express this permutation as $1 \to 6 \to 2 \to 1$, $3 \to 5 \to 3$, $4 \to 4$, or more compactly as $(162)(35)(4)$. This permutation has three *cycles* which could be represented in any number of different ways including $(4)(53)(621)$, but not by $(4)(53)(126)$. To standardize our representations we adopt the following notational convention:

Permutation notation: For a permutation of n elements with k cycles, each cycle begins with its smallest element, and the cycles are listed in increasing order of their smallest element.

Thus the arrangements 265431, 612345, and 134625 are to be represented by the permutations $(162)(35)(4)$, (123456), and $(1)(25643)$ respectively.

So the denominator of H_n counts permutations. We will show that the numerator of H_n is a Stirling number of the first kind which counts special kinds of permutations.

7.2 Stirling Numbers of the First Kind

We begin with a combinatorial definition for $\left[{n \atop k}\right]$, a Stirling number of the first kind.

Definition For integers $n \geq k \geq 0$, let $\left[{n \atop k}\right]$ count the number of permutations of n elements with exactly k cycles.

Equivalently $\left[{n \atop k}\right]$ counts the number of ways for n distinct people to sit around k identical circular tables, where no tables are allowed to be empty. $\left[{n \atop k}\right]$ is called the (unsigned) *Stirling number* of the first kind. As an example, $\left[{3 \atop 2}\right] = 3$ since one person must sit alone at a table and the other two have one way to sit at the other table. That is, $\left[{3 \atop 2}\right]$ counts the three permutations by $(1)(23)$, $(13)(2)$ and $(12)(3)$.

Likewise, $\left[{4 \atop 2}\right] = 11$ counts the eleven permutations $(1)(234)$, $(1)(243)$, $(12)(34)$, $(13)(24)$, $(14)(23)$, $(123)(4)$, $(132)(4)$, $(124)(3)$, $(142)(3)$, $(134)(2)$, $(143)(2)$.

At this point, we have all the tools needed to understand harmonic numbers combinatorially, and the reader who wants to get right to the harmonic numbers can go directly to the next section. In the remainder of this section, we shall combinatorially explore more properties of Stirling numbers of the first kind. Although $\left[{n \atop k}\right]$ has no explicit formula, it satisfies many nice properties.

Since every permutation can be factored into some number of cycles, we immediately have

Identity 182 *For $n \geq 1$,*

$$\sum_{k=1}^{n} \left[{n \atop k}\right] = n!$$

In case the reader is not comfortable with permutations, we present another proof of the above identity using people and tables.

Question: How many ways can n people be seated around n indistinguishable circular tables?

Answer 1: By conditioning on the number of nonempty tables $\sum_{k=1}^{n} \left[{n \atop k}\right]$.

Answer 2: Person 1 sits down at a table (one way, since tables are indistinguishable). Then person 2 has two choices: either sit to the right of person 1 (which is equivalent to sitting to the left) or start a new table. Regardless of 2s decision, person 3 has three choices: either sit to the right of 1, sit to the right of 2, or start a new table. In general for $1 \leq k \leq n$, person k will have k choices: sit to the right of 1 or 2 or $\cdots$ or $(k-1)$, or start a new table. Altogether there are $n!$ possible arrangements.

For a more general version of this seating argument, see the proof of Identity 218 in Chapter 8.

From the definition, we see that $\left[{n \atop n}\right] = 1$ counts the permutation $(1)(2)(3)\cdots(n)$ and $\left[{0 \atop 0}\right] = 1$, but otherwise $\left[{n \atop 0}\right] = 0$. Likewise, we have $\left[{n \atop k}\right] = 0$ whenever $n < 0$ or $k < 0$ or $n < k$. Now $\left[{n \atop n-1}\right] = \binom{n}{2}$ since a permutation with $n-1$ cycles is determined by which two elements appear in the same cycle. Also, for $n \geq 1$,

$$\left[{n \atop 1}\right] = (n-1)!$$

since a permutation of n elements in one cycle must be of the form $(1a_2a_3\cdots a_n)$ where $a_2a_3\cdots a_n$ is an arrangement of the numbers 2 through n.

The notation for Stirling numbers of the first kind is suggestive of binomial coefficients because they share similar properties. As Identity 127 can be used to recursively compute binomial coefficients, the following identity can be used to recursively compute $\left[{n\atop k}\right]$.

Identity 183 *For $n \geq k \geq 1$,*

$$\left[{n\atop k}\right] = \left[{n-1\atop k-1}\right] + (n-1)\left[{n-1\atop k}\right].$$

Question: How many permutations of n elements have exactly k cycles?

Answer 1: By definition, $\left[{n\atop k}\right]$.

Answer 2: Condition on whether or not element n is alone in its own cycle or not. If n is alone in its own cycle, then the remaining $n-1$ elements can be placed into $k-1$ cycles in $\left[{n-1\atop k-1}\right]$ ways. If n is not to be alone, then we first arrange elements 1 through $n-1$ into k cycles (there are $\left[{n-1\atop k}\right]$ ways to do this), then insert element n to the right of any element. This gives us $(n-1)\left[{n-1\atop k}\right]$ total permutations where n is not alone. Altogether, we have $\left[{n-1\atop k-1}\right] + (n-1)\left[{n-1\atop k}\right]$ permutations with k cycles. A concrete example is given in Figure 7.1.

Just as binomial coefficients can be computed by Pascal's triangle, Identity 183 allows Stirling numbers to be displayed in a similar way.

$n \setminus k$	0	1	2	3	4	5	6	7
0	1	0	0	0	0	0	0	0
1	0	1	0	0	0	0	0	0
2	0	1	1	0	0	0	0	0
3	0	2	3	1	0	0	0	0
4	0	6	11	6	1	0	0	0
5	0	24	50	35	10	1	0	0
6	0	120	274	225	85	15	1	0
7	0	720	1764	1624	735	175	21	1

Table 7.1. Stirling numbers of the first kind $\left[{n\atop k}\right]$.

Starting in row 2 of Table 7.1, it appears that $\sum \left[{n\atop k}\right]$ over odd k is the same as when summed over even k. This is not a coincidence as the next identity shows.

Identity 184 *For $n \geq 2$,*

$$\sum_{k=1}^{n}\left[{n\atop k}\right](-1)^k = 0.$$

Set 1: The set of permutations of n elements with an even number of cycles. This set has size $\sum_{k \text{ even}}\left[{n\atop k}\right]$.

(a) 5 is alone in a cycle:

$$\begin{bmatrix} 4 \\ 2 \end{bmatrix} \text{ choices } \left\{ \begin{array}{l} (\mathbf{1})(\mathbf{234})(5) \\ (\mathbf{1})(\mathbf{243})(5) \\ (\mathbf{12})(\mathbf{34})(5) \\ (\mathbf{13})(\mathbf{24})(5) \\ (\mathbf{14})(\mathbf{23})(5) \\ (\mathbf{123})(\mathbf{4})(5) \\ (\mathbf{124})(\mathbf{3})(5) \\ (\mathbf{132})(\mathbf{4})(5) \\ (\mathbf{134})(\mathbf{2})(5) \\ (\mathbf{142})(\mathbf{3})(5) \\ (\mathbf{143})(\mathbf{2})(5) \end{array} \right.$$

(b) 5 is not alone in a cycle:

$$\text{Each of the } \begin{bmatrix} 4 \\ 3 \end{bmatrix} \text{ choices } \left\{ \begin{array}{l} (1)(2)(34) \\ (1)(23)(4) \\ (1)(24)(3) \\ (12)(3)(4) \\ (13)(2)(4) \\ (14)(2)(3) \end{array} \right. \text{ leads to 4 placement choices for 5.}$$

$$\text{For example } (14)(2)(3) \Rightarrow \left\{ \begin{array}{l} (154)(2)(3) \\ (145)(2)(3) \\ (14)(25)(3) \\ (14)(2)(35) \end{array} \right. .$$

Figure 7.1. When considering $\begin{bmatrix} 5 \\ 3 \end{bmatrix}$ either (a) 5 appears alone in a cycle or (b) 5 is immediately to the right of 1, 2, 3, or 4 in one of $\begin{bmatrix} 4 \\ 3 \end{bmatrix}$ arrangements.

Set 2: The set of permutations of n elements with an odd number of cycles. This set has size $\sum_{k \text{ odd}} \begin{bmatrix} n \\ k \end{bmatrix}$.

Correspondence: For any permutation with at least two elements, the element 2 must either appear in the first cycle or as the first element of the second cycle. In the first case, the permutation begins $(1a_1a_2 \cdots a_j 2b_1b_2 \cdots b_k)(c_1c_2 \cdots) \cdots$ where j and k are nonnegative. By converting this to $(1a_1a_2 \cdots a_j)(2b_1b_2 \cdots b_k)(c_1c_2 \cdots) \cdots$, we obtain a permutation with 1 and 2 in different cycles, but with one more cycle than before. Likewise, if 1 and 2 are in different cycles (necessarily leading the first two cycles), then by merging the first two cycles they are now in the same cycle, but with one fewer cycle than before. Thus every permutation corresponds to a unique permutation of opposite parity, as desired.

The Stirling numbers of the first kind are frequently defined as coefficients in the expansion of the rising factorial function [20]:

Identity 185

$$x(x+1)(x+2) \cdots (x+n-1) = \sum_{m=1}^{n} \begin{bmatrix} n \\ m \end{bmatrix} x^m.$$

Just about any Stirling number fact can be proved using this definition. To show that this algebraic definition of Stirling numbers is equivalent to the combinatorial definition, one typically proves that both satisfy the same initial conditions and recurrence relation. However a more direct correspondence exists [2], which we illustrate with an example.

By the algebraic definition, the Stirling number $\left[{10 \atop 3}\right]$ is the coefficient of x^3 in the expansion $x(x+1)(x+2)\cdots(x+9)$. The combinatorial definition says $\left[{10 \atop 3}\right]$ counts the number of ways that elements $0,1,2,\ldots,9$ can sit around three identical circular tables. Why are these definitions the same? The first definition gives the sum of all products of seven numbers chosen from among 1 through 9. Surely this must be counting something. What is a term like $1\cdot2\cdot3\cdot5\cdot6\cdot8\cdot9$ counting? As illustrated in Figure 7.2, this counts the number of ways elements 0 through 9 can seat themselves around three identical tables where the smallest elements of the tables are the "missing" numbers 0, 4, and 7. To see this, we pre-seat numbers $0,4,7$ then seat the remaining numbers one at a time in increasing order. The number 1 has just one option—sit next to 0. The number 2 then has two options—sit to the right of 0 or sit to the right of 1. The number 3 now has three options—sit to the right of 0 or 1 or 2. The number 4 is already seated. Now number 5 has five options—sit to the right of 0 or 1 or 2 or 3 or 4, and so on. A general combinatorial proof of Identity 185 can also be done by the preceding (or should that be "pre-seating"?) argument.

Cartoon by Greg Levin.

Figure 7.2. How many ways can the numbers $1,2,3,5,6,8,9$ seat themselves around these tables?

7.3 Combinatorial Interpretation of Harmonic Numbers

We are now ready to state the main result of this chapter.

Combinatorial Theorem 11 *For $n \geq 0$, the nth harmonic number is*

$$H_n = \frac{\left[{n+1 \atop 2}\right]}{n!}.$$

Before proving this, we set some notational conventions. Let $\mathcal{T}_n$ denote the set of arrangements of the numbers 1 through n into two disjoint, nonempty cycles. Thus $|\mathcal{T}_n| = \left[{n \atop 2}\right]$. For example, $\mathcal{T}_9$ includes the permutation $(185274)(396)$, but not $(195)(2487)(36)$ nor $(123)(4567)(8)(9)$. By our notation for permutations, the cycle containing 1 is always written first; consequently we call it the *left cycle*. The remaining cycle is called the *right cycle*. All permutations in $\mathcal{T}_n$ are of the form $(a_1 a_2 \ldots a_j)(a_{j+1} \ldots a_n)$ where $1 \leq j \leq n-1$, $a_1 = 1$, and a_{j+1} is the smallest element of the right cycle.

Theorem 11 can be rewritten as follows.

Identity 186 *For $n \geq 0$, $\left[{n+1 \atop 2}\right] = n!H_n$.*

Question: How many permutations of $n+1$ elements have exactly 2 cycles?

Answer 1: By definition, $\left[{n+1 \atop 2}\right]$.

Answer 2: Condition on the number of elements in the right cycle. For $1 \leq k \leq n$, we create a permutation of $n+1$ elements with k elements in the right cycle and $n-k+1$ elements in the left cycle. First choose k elements from $\{2, \ldots, n+1\}$ ($\binom{n}{k}$ ways), arrange these elements in the right cycle ($(k-1)!$ ways), then arrange the remaining $n-k$ elements in the left cycle following the number 1 ($(n-k)!$ ways). Hence there are $\binom{n}{k}(k-1)!(n-k)! = \frac{n!}{k}$ permutations of $\mathcal{T}_{n+1}$ with k elements in the right cycle. Summing over all k, the number of permutations of $n+1$ elements with exactly two cycles is $\sum_{k=1}^{n} \frac{n!}{k} = n!H_n$, as desired.

Here is another way to see that $\frac{n!}{k}$ counts permutations with two cycles where the second cycle has k elements. There are $n!$ ways to arrange the numbers 1 through $n+1$ such that the number 1 is written first. Such an arrangement has the form $1a_1 a_2 \cdots a_n$. From this we create the 2-cycled permutation $(1a_1 a_2 \cdots a_{n-k})(a_{n-k+1}a_{n-k+2} \cdots a_n)$ with k elements in the right cycle, but will it be in standard form? Only if a_{n-k+1} is the smallest number in the set $\{a_{n-k+1}, a_{n-k+2}, \ldots, a_n\}$. This happens one time out of k. Hence there are $\frac{n!}{k}$ permutations of $n+1$ elements with exactly k elements in the right cycle.

Using an alternate interpretation of $\frac{n!}{k}$ gives a different method to answer the combinatorial question at hand.

Answer 2′: Condition on the smallest element in the right cycle. For $2 \leq r \leq n+1$, we create a permutation of $n+1$ numbers where the right cycle begins with r. Such a permutation is of the form $(1 \cdots)(r \cdots)$, where elements 1 through $r-1$ all appear in the left cycle, and elements $r+1$ through $n+1$ can go in either cycle. To count this, arrange elements 1 through $r-1$ into the left cycle, listing element 1 first. There are $(r-2)!$ ways to do this. Place element r into the right cycle. Now we insert elements $r + 1$ through $n + 1$, one at a time, each immediately to the *right* of an already placed element. In this way, elements 1 and r remain first (and smallest) in their cycles. Specifically, the element $r + 1$ can go to the right of any of the elements 1 through r. Next, $r + 2$ can go to the right of any of the elements 1 through $r + 1$. Continuing in this way, the number of ways to insert elements $r + 1$ through $n + 1$ is

$$r(r + 1)(r + 2) \cdots n = \frac{n!}{(r - 1)!}.$$

This process creates a permutation in $\mathcal{T}_{n+1}$ with r as the smallest element in the right cycle. Thus, for $2 \le r \le n + 1$, there are

$$(r - 2)! \frac{n!}{(r - 1)!} = \frac{n!}{r - 1}$$

permutations with two cycles, where r is the smallest element in the right cycle. Summing over all values of r gives us

$$\sum_{r=2}^{n+1} \frac{n!}{r - 1} = n! \sum_{k=1}^{n} \frac{1}{k} = n! H_n$$

permutations, as desired.

A final way to see that $n!/(r - 1)$ counts permutations of the form $(1 \cdots)(r \cdots)$ is to list the numbers 1 through $n + 1$ in any order with the provision that 1 be listed first. There are $n!$ ways to do this. We then convert our list $1 \; a_2 \; a_3 \cdots \; r \cdots \; a_{n+1}$ to the permutation $(1 \; a_2 \; a_3 \cdots)(r \cdots a_{n+1})$ by inserting parentheses. This permutation satisfies our notational convention if and only if the number r is listed to the right of elements $2, 3, \ldots, r - 1$. This has probability $1/(r - 1)$ since any of the elements $2, 3, \ldots, r$ have the same chance of being listed last among them. Hence the number of permutations that satisfy our conditions is $n!/(r - 1)$.

7.4   Recounting Harmonic Identities

With our understanding of the interactions between harmonic and Stirling numbers, we now provide combinatorial explanations of other harmonic identities.

In this section, we convert Identities 178, 179, and 180 into statements about Stirling numbers, then explain them combinatorially. Our combinatorial proofs of Theorem 11 were obtained by partitioning the set $\mathcal{T}_{n+1}$ according to the size of the right cycle or the minimum element of the right cycle, respectively. In what follows, we shall transform harmonic Identities 178, 179 and 180 into three Stirling number identities, each with $\left[{n \atop 2} \right]$ on the left-hand side. The right-hand sides will be combinatorially explained by partitioning $\mathcal{T}_n$ according to the location of element 2, the largest of the last t elements, or the "neighborhood" of the elements 1 through m.

Thus for the next three "harmonic" identities, we shall repeat the same question.

Question: How many permutations of n elements have exactly two cycles?

Answer 1: By definition, $\left[{n \atop 2} \right]$.

The challenge will be to find a simple combinatorial interpretation of the right side of the identity.

Our first identity is equivalent to Identity 178. The equivalence can be seen by re-indexing ($n := n - 1$) and applying Combinatorial Theorem 11.

Identity 187 *For $n \geq 2$,*

$$\begin{bmatrix} n \\ 2 \end{bmatrix} = (n-1)! + \sum_{k=1}^{n-2} \frac{(n-2)!}{k!} \begin{bmatrix} k+1 \\ 2 \end{bmatrix}.$$

Answer 2: Here we condition on whether or not the element 2 appears in the left cycle, and if so, how many elements appear to the right of 2. From our second proof of Combinatorial Theorem 11, we know that $(n-1)!$ counts the number of permutations in $\mathcal{T}_n$ where the number 2 appears in the right cycle. It remains to show that

$$\sum_{k=1}^{n-2} \frac{(n-2)!}{k!} \begin{bmatrix} k+1 \\ 2 \end{bmatrix}$$

counts the number of permutations in $\mathcal{T}_n$ where 2 is in the left cycle. Such a permutation is of the form

$$(1 \ a_1 \ a_2 \cdots a_{n-2-k} \ 2 \ b_1 \ b_2 \cdots b_{j-1})(b_j \cdots b_k),$$

for some $1 \leq k \leq n - 2$ and $1 \leq j \leq k$. Next we assert that the number of such permutations with exactly k terms to the right of 2 is $\frac{(n-2)!}{k!} \begin{bmatrix} k+1 \\ 2 \end{bmatrix}$.

To see this, we can select $a_1, a_2, \ldots, a_{n-2-k}$ from the set $\{3, \ldots, n\}$ in $(n-2)!/k!$ ways. From the unchosen elements, there are $\begin{bmatrix} k+1 \\ 2 \end{bmatrix}$ ways to create two nonempty cycles of the form $(2b_1 \ldots b_{j-1})(b_j \ldots b_k)$ where $1 \leq j \leq k$. Hence there are

$$\frac{(n-2)!}{k!} \begin{bmatrix} k+1 \\ 2 \end{bmatrix}$$

permutations in $\mathcal{T}_n$ with exactly k terms to the right of 2, as was to be shown.

We apply a different strategy to prove the more general Identity 179, which after applying Combinatorial Theorem 11 and re-indexing ($n := n - 1$, $m := t - 1$, and $k := k - 2$) is equivalent to the following identity.

Identity 188 *For $1 \leq t \leq n - 1$,*

$$\begin{bmatrix} n \\ 2 \end{bmatrix} = \frac{(n-1)!}{t} + t \sum_{k=t+1}^{n} \begin{bmatrix} k-1 \\ 2 \end{bmatrix} \frac{(n-1-t)!}{(k-1-t)!}.$$

Answer 2: Here we condition on whether or not the *largest of the last t elements* is alone in its own cycle, and if not, the value of the largest of the last t elements. For $1 \leq t \leq n - 1$, we define the *last t elements* of $(1a_2 \cdots a_j)(a_{j+1} \cdots a_n)$ to be the elements $a_n, a_{n-1}, \ldots, a_{n+1-t}$, even if some of them are in the left cycle. For example, the last five elements of $(185274)(396)$ are $6, 9, 3, 4,$ and 7.

Next we claim that for $1 \leq t \leq n - 1$, the number of permutations in $\mathcal{T}_n$ where the largest of the last t elements is alone in the right cycle is $(n-1)!/t$.

Here, we are counting permutations of the form $(1a_2 \ldots a_{n-1})(a_n)$ where $a_n = \max\{a_{n+1-t}, a_{n+2-t}, \ldots, a_{n-1}, a_n\}$. Among all $(n-1)!$ permutations of this form, the largest of the last t elements is equally likely to be anywhere among the last t positions. Hence $(n-1)!/t$ of them have the largest of the last t elements in the last position.

Next we claim that for $1 \leq t \leq n-1$, the number of permutations of $\mathcal{T}_n$ where the largest of the last t elements is *not* alone in the right cycle is

$$t \sum_{k=t+1}^{n} \begin{bmatrix} k-1 \\ 2 \end{bmatrix} \frac{(n-1-t)!}{(k-1-t)!}.$$

To see this, we count the number of such permutations where the largest of the last t elements is equal to k. Since the number 1 is not listed among the last t elements, we have $t+1 \leq k \leq n$. To construct such a permutation, we begin by arranging numbers 1 through $k-1$ into two cycles. Then insert the number k to the right of any of the last t elements. There are $\begin{bmatrix} k-1 \\ 2 \end{bmatrix} t$ ways to do this. The right cycle contains at least one element less than k so k is not alone in the right cycle (and could even be in the left cycle). So that k remains the largest among the last t elements, we insert elements $k+1$ through n, one at a time, to the right of any but the last t elements. There are

$$(k-t)(k+1-t)\cdots(n-1-t) = \frac{(n-1-t)!}{(k-1-t)!}$$

ways to do this. Hence there are

$$t \begin{bmatrix} k-1 \\ 2 \end{bmatrix} \frac{(n-1-t)!}{(k-1-t)!}$$

permutations where the largest of the last t elements equals k, and it is not alone in the right cycle. Summing over all possible values of k gives us

$$t \sum_{k=t+1}^{n} \begin{bmatrix} k-1 \\ 2 \end{bmatrix} \frac{(n-1-t)!}{(k-1-t)!}$$

permutations altogether.

Notice that when $t = 1$, Identity 188 simplifies to Identity 187.

For our final identity, we convert Identity 180 to Stirling numbers using Combinatorial Theorem 11 and re-indexing ($n := n-1$, $m := m-1$, and $k := t-1$.) This gives us

Identity 189 *For $1 \leq m \leq n$,*

$$\begin{bmatrix} n \\ 2 \end{bmatrix} = \begin{bmatrix} m \\ 2 \end{bmatrix} \frac{(n-1)!}{(m-1)!} + \sum_{t=m}^{n-1} \binom{t-1}{m-1} \frac{(m-1)!(n-m)!}{(n-t)}.$$

Answer 2: We condition on whether numbers 1 through m all appear in the left cycle, and if so, how many elements appear in the left cycle. First we claim that for $1 \leq m \leq n$, the number of permutations in $\mathcal{T}_n$ that do not have elements $1, 2, \ldots m$ all in the left cycle is $\begin{bmatrix} m \\ 2 \end{bmatrix} \frac{(n-1)!}{(m-1)!}$. For these permutations, the elements 1 through m can be arranged into two cycles in $\begin{bmatrix} m \\ 2 \end{bmatrix}$ ways. Next insert the remaining elements

$m+1$ through n, one at a time, to the right of any existing element. Hence there are $m(m+1)\cdots(n-1) = (n-1)!/(m-1)!$ ways to insert these elements. Altogether there are

$$\begin{bmatrix} m \\ 2 \end{bmatrix} \frac{(n-1)!}{(m-1)!}$$

such permutations.

Next we show that

$$\sum_{t=m}^{n-1} \binom{t-1}{m-1} \frac{(m-1)!(n-m)!}{(n-t)}$$

counts the number of permutations in $\mathcal{T}_n$ where elements 1 through m are all in the left cycle. To see this, we claim that for $m \le t \le n-1$, the summand counts the permutations described above with exactly t elements in the left cycle and $n-t$ elements in the right cycle. To create such a permutation, we first place the number 1 at the front of the left cycle. Now choose $m-1$ of the remaining $t-1$ spots in the left cycle to be assigned the elements $\{2,\ldots,m\}$. There are $\binom{t-1}{m-1}$ ways to select these $m-1$ spots and $(m-1)!$ ways to arrange elements $2,\ldots,m-1$ in those spots. For example, to guarantee that elements $1,2,3,4$ appear in the left cycle of Figure 7.3, we select three of the five open spots in which to arrange $2,3,4$. The insertion of $5,6,7,8,9$ remains.

Now there are $(n-m)!$ ways to arrange elements $m+1$ through n in the remaining spots, but only $\frac{1}{n-t}$th of them will put the smallest element of the right cycle at the front of the right cycle. Hence, elements $m+1$ through n can be arranged in $\frac{(n-m)!}{n-t}$ legal ways. Altogether there are

$$\binom{t-1}{m-1}(m-1)!\frac{(n-m)!}{(n-t)}$$

ways to satisfy our conditions, as desired.

An application of harmonic numbers arises in calculating the average number of cycles in a permutation of n elements. Specifically,

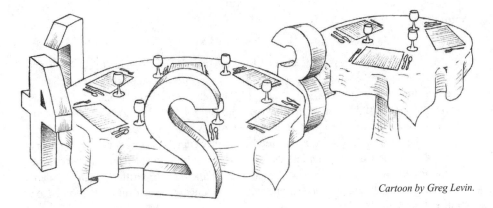

Cartoon by Greg Levin.

Figure 7.3. In $\mathcal{T}_9$, a permutation with $1,2,3,4$ in a left cycle containing exactly six elements is created by first selecting three of the five open spots, and then arranging $2,3,4$ in them. Subsequently, $5,6,7,8,9$ will be arranged in the remaining spots.

Theorem 12 *On average, a permutation of n elements has H_n cycles.*

There are $n!$ permutations of n elements, among which $\left[{n \atop k}\right]$ of them have k cycles. Consequently, Theorem 12 says

$$\frac{\sum_{k=1}^{n} k \left[{n \atop k}\right]}{n!} = H_n, \tag{7.1}$$

or equivalently, by Combinatorial Theorem 11,

Identity 190 *For $n \geq 1$,*

$$\sum_{k=1}^{n} k \left[{n \atop k}\right] = \left[{n+1 \atop 2}\right].$$

Set 1: The permutations of $\{1, \ldots, n\}$ with an arbitrary number of cycles, where one of the cycles is distinguished in some way. For example $\underline{(1284)}(365)(79)$, $(1284)\underline{(365)}(79)$, and $(1284)(365)\underline{(79)}$ are three different arrangements with $k = 3$. Since a permutation with k cycles leads to k distinguished permutations, this set has size equal to $\sum_{k=1}^{n} k \left[{n \atop k}\right]$.

Set 2: The set of permutations of $\{0, 1, \ldots, n\}$ with exactly two cycles. This set has size $\left[{n+1 \atop 2}\right]$.

Correspondence: We illustrate our one-to-one correspondence between these two sets by the following three examples.

$$\underline{(1284)}(365)(79) \Longleftrightarrow (079365)(1284)$$
$$(1284)\underline{(365)}(79) \Longleftrightarrow (0791284)(365)$$
$$(1284)(365)\underline{(79)} \Longleftrightarrow (03651284)(79)$$

In general, we transform the permutation with n elements

$$(C_k)(C_{k-1}) \cdots (C_{j+1})\underline{(C_j)}(C_{j-1}) \cdots (C_2)(C_1)$$

into

$$(0 \ C_1 \ C_2 \cdots C_{j-1} \ C_{j+1} \cdots C_{k-1} \ C_k)(C_j).$$

The process is easily reversed. Given $(0 \ a_1 \cdots a_{n-j})(b_1 \cdots b_j)$ in $\mathcal{T}_{n+1}$, the right cycle becomes the distinguished cycle $(b_1 \cdots b_j)$. The distinguished cycle is then inserted among the cycles $C_{k-1}, \ldots C_2, \overline{C_1}$, which are generated one at a time as follows: C_1 (the rightmost cycle) begins with a_1 followed by a_2 and so on until we encounter a number a_i that is less than a_1. Assuming such an a_i exists (i.e., $a_1 \neq 1$), begin cycle C_2 with a_i and repeat the procedure, starting a new cycle every time we encounter a new smallest element. The resulting cycles (after inserting the distinguished one in its proper place) will be an element of $\mathcal{T}_n$ written in our standard notation. Hence we have a one-to-one correspondence between the sets counted on both sides of Identity 190.

By modifying this procedure, as in the exercises, other relationships can be derived as well.

7.5 Stirling Numbers of the Second Kind

It would not be fair to discuss Stirling numbers of the first kind without saying a few words about their intimate companions, the Stirling numbers of the second kind.

> **Definition:** For integers n, k, let $\left\{ {n \atop k} \right\}$ count the number of ways to partition a set with n elements into k disjoint, nonempty subsets. $\left\{ {n \atop k} \right\}$ is called a Stirling number of the second kind.

For example, $\left\{ {4 \atop 2} \right\} = 7$ counts the partitions of $\{1, 2, 3, 4\}$ into two subsets, namely: $\{1\}\{2, 3, 4\}$, $\{1, 2\}\{3, 4\}$, $\{1, 3\}\{2, 4\}$, $\{1, 4\}\{2, 3\}$, $\{1, 2, 3\}\{4\}$, $\{1, 2, 4\}\{3\}$, $\{1, 3, 4\}\{2\}$.

Notice that the partition $\{1, 3\}\{2, 4\}$ is the same as the partition $\{2, 4\}\{3, 1\}$. Hence we adopt the following convention.

> **Set partition notation:** For a partition of $\{1, \ldots, n\}$ into k disjoint subsets, each subset is written in increasing order of its elements, and the subsets are listed in increasing order of their smallest element.

The seven partitions of $\{1, 2, 3, 4\}$ into two subsets given previously are written using this convention.

Just as with Stirling numbers of the first kind, $\left\{ {n \atop n} \right\} = 1$ counts the partition

$$\{1\}\{2\}\{3\} \cdots \{n\},$$

and $\left\{ {0 \atop 0} \right\} = 1$, but otherwise $\left\{ {n \atop 0} \right\} = 0$. Likewise, we have $\left\{ {n \atop k} \right\} = 0$ whenever $n < 0$, or $k < 0$, or $n < k$. Some values of $\left\{ {n \atop k} \right\}$ can be computed straight from the definition. For example, for $n \geq 1$, $\left\{ {n \atop 1} \right\} = 1$ counts the partition $\{1, \ldots, n\}$ and $\left\{ {n \atop 2} \right\} = 2^{n-1} - 1$, since the right subset is a nonempty subset of $\{2, \ldots, n\}$, and all the other elements appear in the left subset with 1. Also, $\left\{ {n \atop n-1} \right\} = \binom{n}{2}$ since a partition with $n - 1$ subsets is determined by which two elements appear together in a subset.

As with binomial coefficients (Identity 127) and Stirling numbers of the first kind (Identity 183), $\left\{ {n \atop k} \right\}$ can be computed recursively.

Identity 191 *For $n \geq k \geq 1$,*

$$\left\{ {n \atop k} \right\} = \left\{ {n-1 \atop k-1} \right\} + k \left\{ {n-1 \atop k} \right\}.$$

Question: How many partitions of an n-element set have exactly k subsets?

Answer 1: By definition, $\left\{ {n \atop k} \right\}$.

Answer 2: Condition on whether element n is alone in its own subset or not. If n is alone in its own subset, then the remaining $n - 1$ elements can be placed into $k - 1$ subsets in $\left\{ {n-1 \atop k-1} \right\}$ ways. If n is not to be alone, then we first arrange elements 1 through $n - 1$ into k subsets (there are $\left\{ {n-1 \atop k} \right\}$ ways to do this), then insert element n into any of the k subsets. This gives us $k \left\{ {n-1 \atop k} \right\}$ total partitions where n is not alone. Altogether, we have $\left\{ {n-1 \atop k-1} \right\} + k \left\{ {n-1 \atop k} \right\}$ partitions with k subsets.

Using this identity, we can easily compute Stirling numbers of the second kind. See Table 7.2.

$n \setminus k$	0	1	2	3	4	5	6	7
0	1	0	0	0	0	0	0	0
1	0	1	0	0	0	0	0	0
2	0	1	1	0	0	0	0	0
3	0	1	3	1	0	0	0	0
4	0	1	7	6	1	0	0	0
5	0	1	15	25	10	1	0	0
6	0	1	31	90	65	15	1	0
7	0	1	63	301	350	140	21	1

Table 7.2. Stirling numbers of the second kind $\left\{ {n \atop k} \right\}$.

As with the other Stirling numbers, Stirling numbers of the second kind also have an algebraic definition that can be explained by a combinatorial argument. For $x, k \geq 0$, define the falling factorial function $x_{(k)}$ to be the product of the k consecutive integers beginning with x and decreasing. Thus $x_{(0)} = 1$, $x_{(1)} = x$, $x_{(2)} = x(x-1)$, and in general for $k \geq 1$,

$$x_{(k)} = x(x-1)(x-2)\cdots(x-k+1).$$

$x_{(k)}$ can also be thought of as a polynomial in x.

Identity 192 *For $n \geq 0$,*

$$x^n = \sum_{k=0}^{n} \left\{ {n \atop k} \right\} x_{(k)}.$$

Question: How many ways can n students be assigned to x different classrooms, where some classrooms are allowed to be empty?

Answer 1: Since each student can be assigned one of x classrooms, there are x^n such assignments.

Answer 2: Condition on the number of occupied classrooms. For $0 \leq k \leq n$, there are $\left\{ {n \atop k} \right\}$ ways to partition the students into k nonempty subsets, followed by $x_{(k)}$ ways to assign each of these subsets to a classroom.

There is not a simple closed form for $\sum_{k=0}^{n} \left\{ {n \atop k} \right\} x^k$, but if we instead fix k and let n vary, then we have the following generating function identity.

Identity 193 *For $k \geq 0$, and for all x sufficiently small,*

$$\sum_{n \geq 0} \left\{ {n \atop k} \right\} x^n = \frac{x^k}{(1-x)(1-2x)(1-3x)\cdots(1-kx)}.$$

We know that the expression $\frac{1}{1-jx}$ is equal to the geometric series $1 + jx + j^2 x^2 + j^3 x^3 + \cdots$ for all sufficiently small values of x. Hence the identity asserts that $\left\{ {n \atop k} \right\}$ is the x^{n-k} coefficient of

$$(1 + x + x^2 + x^3 + \cdots)(1 + 2x + 4x^2 + 8x^3 + \cdots)\cdots(1 + kx + k^2 x^2 + k^3 x^3 + \cdots).$$

For instance when $n = 20$ and $k = 3$, $\left\{ {20 \atop 3} \right\}$ is the coefficient of the x^{17} term of

$$(1 + x + x^2 + x^3 + \cdots)(1 + 2x + 4x^2 + 8x^3 + \cdots)(1 + 3x + 9x^2 + 27x^3 + \cdots),$$

which is the sum of terms like $(1x^5)(2^8x^8)(3^4x^4)$. What are numbers like $1^52^83^4$ counting? We know that $\left\{{20 \atop 3}\right\}$ counts the number of partitions of the set $\{1, 2, \ldots, 20\}$ into three subsets $\{a_1, a_2, \ldots, \}\{b_1, b_2, \ldots\}\{c_1, c_2, \ldots\}$ where each set is written in increasing order of its elements and $1 = a_1 < b_1 < c_1$. We claim that the number of such partitions with $a_1 = 1, b_1 = 7, c_1 = 16$ is $1^52^83^4$ since elements 2 through 6 must be in the first subset, elements 8 through 15 can be in the first or second subset, and elements 17 through 20 can each be in the first or second or third subset. In general, the x^{n-k} coefficient is the sum of products of the form $1^{e_1}2^{e_2}\cdots k^{e_k}$, where $e_1+e_2+\cdots+e_k = n-k$ which count partitions whose minimum elements are $a_1 = 1, b_1 = 2+e_1, c_1 = 3+e_1+e_2, d_1 = 4+e_1+e_2+e_3$, and so on, up to the kth set with minimum element $k + e_1 + e_2 + \cdots + e_{k-1} = n - e_k$. Once the minimum elements are chosen, the numbers between 1 and b_1 are forced to go into the first subset, the numbers between b_1 and c_1 have two choices, the numbers between c_1 and d_1 have three choices, and so on.

We conclude this section by generalizing Identity 184, combinatorially proving that Stirling numbers of the first and second kind are intimately connected.

Identity 194 *For $m, n \geq 0$,*

$$\sum_{k=0}^{n} \left[{n \atop k}\right]\left\{{k \atop m}\right\}(-1)^k = (-1)^n\delta_{m,n}, \tag{7.2}$$

where $\delta_{m,n} = 1$ if $m = n$, and is 0 otherwise.

Set 1: Let $\mathcal{E}$ be the set of arrangements of n distinct students around an even number of identical circular nonempty tables which are placed into m indistinguishable nonempty classrooms. $|\mathcal{E}| = \sum_{k \text{ even}} \left[{n \atop k}\right]\left\{{k \atop m}\right\}$.

Set 2: Let $\mathcal{O}$ be the set of arrangements of n distinct students around an odd number of identical circular nonempty tables which are placed into m indistinguishable nonempty classrooms. $|\mathcal{O}| = \sum_{k \text{ odd}} \left[{n \atop k}\right]\left\{{k \atop m}\right\}$.

Correspondence: We establish a one-to-one correspondence between $\mathcal{E}$ and $\mathcal{O}$ except for the case where $m = n$. When $m = n$ there is only one way that n students can ultimately be placed into n classrooms and that is for each classroom to contain one table, and each table to contain one student at it. When n is even, $|\mathcal{E}| = 1$ and $|\mathcal{O}| = 0$; when n is odd $|\mathcal{E}| = 0$ and $|\mathcal{O}| = 1$, as predicted. Either way, $|\mathcal{E}| - |\mathcal{O}| = (-1)^n$, as predicted.

When $m > n$, both $\mathcal{E}$ and $\mathcal{O}$ are empty. When $m < n$, every element of $\mathcal{E}$ (or $\mathcal{O}$) must have a classroom with at least two students in it. Now for any element of $\mathcal{E}$, let a be the smallest numbered student who is not in a room by herself, and let b be the next smallest numbered student in the same room. We adopt the convention of placing each table in a given room with the smallest element listed first. If a and b are not at the same table, then our room is a set of tables: $\{(a\cdots), (b\cdots), (c\cdots), \cdots\}$. We transform this (by removing just two parentheses, but leaving everything else listed in the same order) into $\{(a\cdots b\cdots), (c\cdots), \cdots\}$ where a and b are now at the same table. This new allocation has one fewer table than the original one, and is therefore in $\mathcal{O}$. Conversely, if a and b are at the same table, then we can undo this transformation (by reinserting the parentheses) so that they are at different tables, again creating an element of $\mathcal{O}$.

In a similar fashion, it can also be proved that

Identity 195 *For* $m, n \geq 0$,

$$\sum_{k=0}^{n} \left\{ {n \atop k} \right\} \left[{k \atop m} \right] (-1)^k = (-1)^n \delta_{m,n}. \tag{7.3}$$

We leave its proof as an exercise for the reader.

7.6 Notes

Some of the material in this chapter is based on [6], developed with undergraduate Greg Preston. The proof of Identity 185 was first shown to us by Robert Beals [2]. Identities 198 through 210 are listed in [28] as useful Stirling number identities. The final exercise was developed by undergraduates David Gaebler and Robert Gaebler. We thank Greg Levin for the cartoons.

7.7 Exercises

1. Here is an alternative proof of Combinatorial Theorem 11, based on $H_n = H_{n-1} + \frac{1}{n}$. Write $H_n = \frac{a_n}{n!}$ and prove that $a_n = \left\{ {n+1 \atop 2} \right\}$ by showing that they satisfy the same recurrence and initial conditions.

2. Modify the proof of Identity 190 to prove identities:

    Identity 196 *For* $m, n \geq 0$, $\sum_{k=m}^{n} \left[{n \atop k} \right] \binom{k}{m} = \left[{n+1 \atop m+1} \right]$.

    Identity 197 *For* $m, n \geq 0$, $\sum_{k=m}^{n} \left[{n \atop k} \right] 2^k = (n+1)!$

Find combinatorial proofs for the following identities.

Identity 198 *For* $m, n \geq 0$, $\sum_{k=m}^{n} \binom{n}{k} \left\{ {k \atop m} \right\} = \left\{ {n+1 \atop m+1} \right\}$.

Identity 199 *For* $m, n \geq 0$, $\sum_{k=0}^{m} k \left\{ {n+k \atop k} \right\} = \left\{ {m+n+1 \atop m} \right\}$.

Identity 200 *For* $m, n \geq 0$, $\sum_{k=0}^{m} (n+k) \left[{n+k \atop k} \right] = \left[{m+n+1 \atop m} \right]$.

Identity 201 *For* $m, n \geq 0$, $\sum_{k=0}^{n} \left\{ {k \atop m} \right\} (m+1)^{n-k} = \left\{ {n+1 \atop m+1} \right\}$.

Identity 202 *For* $m, n \geq 0$, $\sum_{k=m}^{n} \left[{k \atop m} \right] n!/k! = \left[{n+1 \atop m+1} \right]$.

Identity 203 *For* $1 \leq m \leq n$, $\sum_{k=m}^{n} \left[{n \atop k} \right] \left\{ {k \atop m} \right\} = \binom{n}{m}(n-1)!/(m-1)!$

Identity 204 *For* $\ell, m, n \geq 0$, $\sum_{k=0}^{n} \binom{n}{k} \left\{ {k \atop \ell} \right\} \left\{ {n-k \atop m} \right\} = \left\{ {n \atop \ell+m} \right\} \binom{\ell+m}{\ell}$.

Identity 205 *For* $\ell, m, n \geq 0$, $\sum_{k=0}^{n} \binom{n}{k} \left[{k \atop \ell} \right] \left[{n-k \atop m} \right] = \left[{n \atop \ell+m} \right] \binom{\ell+m}{\ell}$.

Identity 206 *For* $m, n \geq 0$, $\sum_{k=m}^{n} \left\{ {n+1 \atop k+1} \right\} \left[{k \atop m} \right] (-1)^k = (-1)^m \binom{n}{m}$.

The following identities can be proved by finding almost one-to-one correspondences between even sets and odd sets.

Identity 207 *For* $0 \le m \le n$, $\sum_{k=m}^{n} \left[{n+1 \atop k+1} \right] \left\{ {k \atop m} \right\} (-1)^k = n!/m!$

Identity 208 *For* $0 \le m \le n$, $\sum_{k=m}^{n} \binom{n}{k} \left\{ {k+1 \atop m+1} \right\} (-1)^k = (-1)^n \left\{ {n \atop m} \right\}$.

Identity 209 *For* $0 \le m \le n$, $\sum_{k=m}^{n} \left[{n+1 \atop k+1} \right] \binom{k}{m} (-1)^k = (-1)^m \left[{n \atop m} \right]$.

Identity 210 *For* $m, n \ge 0$, $\sum_{k=0}^{m} \binom{m}{k} k^n (-1)^k = (-1)^m m! \left\{ {n \atop m} \right\}$.

Challenging Exercise Conway and Guy define the r-harmonic numbers as follows: Let $H_n^r = 0$ for $r < 0$ or $n \le 0$, $H_n^0 = \frac{1}{n}$ for $n \ge 1$, and for $r, n \ge 1$, let

$$H_n^r = \sum_{i=1}^{n} H_i^{r-1}.$$

Note that H_n^1 is equal to the ordinary harmonic number H_n. Broder [16] defines the r-Stirling number $\left[{n \atop k} \right]_r$ to be the number of permutations of n elements into k cycles where elements 1 through r must all be in different cycles. Prove the following generalization of Combinatorial Theorem 11.

$$H_n^r = \frac{\left[{n+r \atop r+1} \right]_r}{n!}.$$

CHAPTER 8

Number Theory

In this chapter, we have collected identities from arithmetic, algebra and number theory.

8.1 Arithmetic Identities

What could be simpler than the sum of the first n numbers? You probably already know this first identity. In fact, it's a special case of Identity 135 in Chapter 5. You may recall that

$$\sum_{k=1}^{n} k = \frac{n(n+1)}{2}.$$

Combinatorially, this can be rewritten and explained in two different ways. The subsequent identities and their counting proofs hinge on the interpretation of $\frac{n(n+1)}{2}$ as $\binom{n+1}{2}$, a selection without repetition, or $\left(\binom{n}{2}\right)$, a selection with repetition.

Identity 211 *For $n \geq 0$,*

$$\sum_{k=1}^{n} k = \binom{n+1}{2}.$$

> **Question:** How many ways can two different numbers be selected from the set $\{0, 1, \ldots, n\}$?
>
> **Answer 1:** By definition, $\binom{n+1}{2}$.
>
> **Answer 2:** Condition on the larger of the two elements selected. If the larger number is k, the smaller element can be chosen from any of the k members in $\{0, 1, 2, \ldots, k-1\}$. Hence the total number of selections is $\sum_{k=1}^{n} k$.

The next identity uses the alternate interpretation of $\frac{n(n+1)}{2}$ as $\left(\binom{n}{2}\right)$ allowing repetition. Consequently, the question to be answered involves selecting elements from a different set.

Identity 212 *For $n \geq 0$,*

$$\sum_{k=1}^{n} k = \left(\binom{n}{2}\right).$$

Question: How many ways can two numbers be selected from the set $\{1,\dots,n\}$ where repetition of elements is allowed.

Answer 1: By definition, $\left(\binom{n}{2}\right)$.

Answer 2: Condition on the larger of the two elements selected. If the larger number is k, the smaller element can be chosen from any of the k members in $\{1, 2, \dots, k\}$. Hence the number of selections is $\sum_{k=1}^{n} k$.

The beautiful fact that the sum of the cubes of first n numbers is the square of the sum of the first n numbers can likewise be demonstrated two different ways. In each case, we find a one-to-one correspondence between sets $\mathcal{S}$ and $\mathcal{T}$ where $|\mathcal{S}| = \sum_{k=1}^{n} k^3$, and $|\mathcal{T}| = (\frac{n(n+1)}{2})^2$.

Identity 213 *For $n \geq 0$,*

$$\sum_{k=1}^{n} k^3 = \binom{n+1}{2}^2.$$

Set 1: Let $\mathcal{S}$ denote the set of 4-tuples of integers from 0 to n whose last component is strictly bigger than the others; that is,

$$\mathcal{S} = \{(h, i, j, k) \mid 0 \leq h, i, j < k \leq n\}.$$

For $1 \leq k \leq n$, there are k^3 ways to choose h, i, j given the last component k. Hence, $|\mathcal{S}| = \sum_{k=1}^{n} k^3$.

Set 2: Let $\mathcal{T}$ denote the set of ordered pairs of two element subsets of $\{0, \dots, n\}$. If we write the elements of our subsets in increasing order, $\mathcal{T}$ may be expressed as

$$\mathcal{T} = \{(\{x_1, x_2\}, \{x_3, x_4\}) \mid 0 \leq x_1 < x_2 \leq n, \ 0 \leq x_3 < x_4 \leq n\}.$$

Clearly $|\mathcal{T}| = \binom{n+1}{2}^2$.

Correspondence: To see that $\mathcal{S}$ and $\mathcal{T}$ have the same size, we find a one-to-one correspondence $f : \mathcal{S} \to \mathcal{T}$ between these sets. Specifically, let

$$f((h, i, j, k)) = \begin{cases} (\{h, i\}, \{j, k\}) & \text{if } h < i, \\ (\{j, k\}, \{i, h\}) & \text{if } h > i, \\ (\{i, k\}, \{j, k\}) & \text{if } h = i. \end{cases}$$

For example,

$$f((1, 2, 3, 4)) = (\{1, 2\}, \{3, 4\}),$$
$$f((2, 1, 3, 4)) = (\{3, 4\}, \{1, 2\}), \text{ and}$$
$$f((1, 1, 2, 4)) = (\{1, 4\}, \{2, 4\}).$$

Note that f is easily reversed since the cases where $h < i$, $h > i$, and $h = i$ are mapped onto ordered pairs $(\{x_1, x_2\}, \{x_3, x_4\})$ where $x_2 < x_4$, $x_2 > x_4$, and $x_2 = x_4$, respectively.

A simpler correspondence arises when we allow repetition.

Identity 214 *For $n \geq 0$,*

$$\sum_{k=1}^{n} k^3 = \left(\binom{n}{2}\right)^2.$$

Set 1: This time, we let S denote the set of 4-tuples of integers from 1 to n whose last component is *greater than or equal* to the others. Specifically,

$$S = \{(h, i, j, k) \mid 1 \leq h, i, j \leq k \leq n\}.$$

For a given value of k between 1 and n there are k^3 ways to select h, i, and j. Thus $|S| = \sum_{k=1}^{n} k^3$.

Set 2: Let T denote the set of ordered pairs of 2-element multisubsets of $\{1, 2, \ldots, n\}$. Then

$$T = \{(\{x_1, x_2\}, \{x_3, x_4\}) \mid 1 \leq x_1 \leq x_2 \leq n, \ 1 \leq x_3 \leq x_4 \leq n\},$$

and $|T| = \left(\binom{n}{2}\right)^2$.

Correspondence: Here, our bijection $g : S \to T$ has just two cases:

$$g((h, i, j, k)) = \begin{cases} (\{h, i\}, \{j, k\}) & \text{if } h \leq i, \\ (\{j, k\}, \{i, h - 1\}) & \text{if } h > i. \end{cases}$$

For example,

$$g((1, 2, 3, 4)) = (\{1, 2\}, \{3, 4\}),$$
$$g((2, 1, 3, 4)) = (\{3, 4\}, \{1, 1\}).$$

The first case maps onto those $(\{x_1, x_2\}, \{x_3, x_4\})$ where $x_2 \leq x_4$, and the second case maps onto those where $x_2 > x_4$. Hence g is easily reversed.

Another combinatorial approach to this identity is utilized in [34] and [56] using the set S from our first proof. By conditioning on the number of 4-tuples in S with $2, 3$ or 4 distinct elements, it follows that

$$\sum_{k=1}^{n} k^3 = \binom{n+1}{2} + \binom{n+1}{3}6 + \binom{n+1}{4}3!,$$

which algebraically simplifies to

$$\frac{n^2(n+1)^2}{4}.$$

Our proofs of Identities 213 and 214 avoid the use of algebra and derive $\binom{n+1}{2}^2$ in a purely combinatorial way.

Another well-known identity is

$$\sum_{k=1}^{n} k^2 = \frac{n(n+1)(2n+1)}{6} = \frac{1}{4}\binom{2n+2}{3}.$$

By letting $\mathcal{S} = \{(i,j,k) | 0 \leq i,j < k \leq n\}$ and conditioning on the number of elements in $\mathcal{S}$ with two or three distinct elements, we get

$$\sum_{k=1}^{n} k^2 = \binom{n+1}{2} + 2\binom{n+1}{3}$$

which algebraically simplifies to $\frac{n(n+1)(2n+1)}{6}$. The next identity shows how to avoid the algebraic step above entirely. Another proof that "chooses" instead of "multi-chooses" is given in the exercises.

Identity 215 *For $n \geq 0$,*

$$\sum_{k=1}^{n} k^2 = \frac{1}{4}\left(\binom{2n}{3}\right).$$

Set 1: Let $\mathcal{S}$ be the set of 3-tuples of integers from 1 to n whose last component is greater than or equal to the others. That is,

$$\mathcal{S} = \{(i,j,k) | 1 \leq i,j \leq k \leq n\}.$$

So $|\mathcal{S}| = \sum_{k=1}^{n} k^2$.

Set 2: Let $\mathcal{T}$ be the set of 3-element multisets of $\{1,2,\ldots,2n\}$.

$$\mathcal{T} = \{\{x_1, x_2, x_3\} | 1 \leq x_1 \leq x_2 \leq x_3 \leq 2n\},$$

and $|\mathcal{T}| = \left(\binom{2n}{3}\right)$.

Correspondence: Here we provide a 1-to-4 correspondence between $\mathcal{S}$ and $\mathcal{T}$ by sending the element (i,j,k) to four destinations in $\mathcal{T}$. If $i \leq j$, then (i,j,k) is mapped to $\{2i, 2j, 2k\}$ and $\{2i-1, 2j, 2k\}$ and $\{2i-1, 2j-1, 2k\}$ and $\{2i-1, 2j-1, 2k-1\}$. Whereas if $i > j$, then (i,j,k) is mapped to $\{2j, 2i-1, 2k-1\}$ and $\{2j, 2i-1, 2k\}$ and $\{2j, 2i-2, 2k-1\}$ and $\{2j-1, 2i-2, 2k-1\}$. For example, $(1,2,3)$ is mapped to $\{2,4,6\}$, $\{1,4,6\}$, $\{1,3,6\}$, $\{1,3,5\}$. Whereas $(2,1,3)$ is mapped to $\{2,3,5\}$, $\{2,3,6\}$, $\{2,2,5\}$, $\{1,2,5\}$. The mapping is easily reversed by examining the parity of each component of $\{x_1, x_2, x_3\}$. See Figure 8.1.

$$\text{If } i \leq j \text{ then } (i,j,k) \rightarrow \begin{cases} \{2i, & 2j, & 2k\} & eee \\ \{2i-1, & 2j, & 2k\} & oee \\ \{2i-1, & 2j-1, & 2k\} & ooe \\ \{2i-1, & 2j-1, & 2k-1\} & ooo \end{cases}$$

$$\text{If } i > j \text{ then } (i,j,k) \rightarrow \begin{cases} \{2j, & 2i-1, & 2k-1\} & eoo \\ \{2j, & 2i-1, & 2k\} & eoe \\ \{2j, & 2i-2, & 2k-1\} & eeo \\ \{2j-1, & 2i-2, & 2k-1\} & oeo \end{cases}$$

Figure 8.1. Every 3-tuple of integers from $\{1,2,\ldots,n\}$ whose maximum occurs in the last component leads to 4 unique 3-element multisets of $\{1,2,\ldots,2n\}$.

Another useful identity is the finite geometric series which states that for any (real or complex) number $x \neq 1$ and any positive integer n,

$$1 + x + x^2 + x^3 + \cdots + x^{n-1} = \frac{1 - x^n}{1 - x}.$$

We prove this for positive integers x in the following form.

Identity 216 *For $n \geq 1$,*

$$(x - 1)(1 + x + x^2 + \cdots + x^{n-1}) = x^n - 1.$$

Question: How many sequences of n numbers can be created where each number comes from the set $\{1, \ldots, x\}$, but we exclude the sequence consisting of all xs.

Answer 1: $x^n - 1$.

Answer 2: Condition on the location of the first term in the sequence that is not an x. Suppose for $0 \leq k \leq n - 1$, the first non-x occurs at the $(n - k)$th term. Then there are $(x - 1)$ choices for that term, and x^k ways to finish the sequence, yielding $(x - 1)x^k$ such sequences. Consequently, there are $(x - 1)\sum_{k=0}^{n-1} x^k$ such sequences.

Notice that both sides of Identity 216 are polynomials of degree n. Both polynomials agree on more than $n + 1$ inputs (since they agree on all positive integers), therefore they must be equal on all real or complex input values for x. The same is true for the next identity.

Identity 217 *For $n \geq 0$,*

$$\sum_{k \geq 0} \binom{n}{2k} x^{n-2k} = \frac{1}{2}\left[(x + 1)^n + (x - 1)^n\right].$$

Set 1: Let $\mathcal{S}$ be the set of all sequences of n numbers, each chosen from the set $\{1, \ldots, x + 1\}$, where the number $x + 1$ appears an even number of times. The number of sequences where $x + 1$ occurs $2k$ times is $\binom{n}{2k}x^{n-2k}$, since we must first choose which terms are $x + 1$ and the remaining $n - 2k$ terms each have x choices. Consequently, $|\mathcal{S}| = \sum_{k \geq 0} \binom{n}{2k}x^{n-2k}$.

Set 2: Let $\mathcal{R}$ be the set of all sequences of n numbers, each chosen from the set $\{1, \ldots, x+1\}$, where all the numbers are painted red. Let $\mathcal{B}$ be the set of all sequences of n numbers, each chosen from the set $\{1, \ldots, x - 1\}$, where all the numbers are painted blue. Here $|\mathcal{R} \cup \mathcal{B}| = (x + 1)^n + (x - 1)^n$.

Correspondence: We find a 1-to-2 correspondence between $\mathcal{S}$ and $\mathcal{R} \cup \mathcal{B}$. Let $X \in \mathcal{S}$. Then we associate with X the elements X' and X'' in $\mathcal{R} \cup \mathcal{B}$ where X' is the same as X painted red. Thus X' is in $\mathcal{R}$, and will contain an *even* number of $x + 1$'s. X'' depends on whether or not the numbers x or $x + 1$ occur in X. If neither of these number occur in X, then X'' is the same as X painted blue. Here X'' is in $\mathcal{B}$. Otherwise, X has at least one number equal to x or $x + 1$. Let X'' be the same as X painted red, but with the first occurrence of x or $x + 1$ replaced with the other number. Hence X'' is in $\mathcal{R}$ and will contain an *odd* number of $x + 1$s.

The next identity is typically proved by telescoping sums

$$\sum_{k=2}^{n} \frac{1}{k(k-1)} = 1 - \frac{1}{n}.$$

But multiplying through by $n!$ gives us a permutation identity which we will count using the seating argument presented for Identity 182.

Identity 218 *For $n \geq 1$,*

$$\sum_{k=2}^{n} \frac{n!}{k(k-1)} = n! - (n-1)!$$

Question: How many permutations of the numbers 1 through n consist of two or more cycles?

Answer 1: Since there are $(n-1)!$ permutations with just one cycle, there are $n! - (n-1)!$ such permutations.

Answer 2: Condition on the first element of the *second* cycle. For $2 \leq k \leq n$, how many permutations have the number k as the first element of the second cycle? Using the cycle notation set on page 92, recall that the first cycle begins with 1 and must also contain the numbers 2 through $k-1$. Inserting elements one at a time, the numbers 2 through $k-1$ can be arranged $(k-2)!$ ways in the first cycle. The number k is required to lead the second cycle. Now the number $k+1$ can be inserted $k+1$ places: to the right of any of the existing k elements or *it may start a new cycle*. Likewise the number $k+2$ can be inserted $k+2$ places, and so on. Hence the number of such permutations is $(k-2)!(k+1)(k+2)\cdots n = \frac{n!}{k(k-1)}$. Altogether, there are

$$\sum_{k=2}^{n} \frac{n!}{k(k-1)}$$

permutations with two or more cycles.

8.2   Algebra and Number Theory

Counting arguments can be profitably used in abstract algebra and number theory. Entire books have been written on the subject (e.g., [24]). Here we merely give the flavor of what can be done.

Theorem 13 *If p is prime, then p divides $\binom{p}{k}$ for $0 < k < p$.*

Proof 1. By Identity 130 of Chapter 5, $k\binom{p}{k} = p\binom{p-1}{k-1}$ is a multiple of p. Since p is prime, it has no common prime factors with k. Hence p must divide $\binom{p}{k}$.   ◇

Our second proof of this theorem is more in the spirit of the proofs that follow.

Definition Let S be a set and g be a function from S to S. For x in S, define the *orbit* of x to be the set $\{x, g(x), g^{(2)}(x), g^{(3)}(x), \ldots\}$. Here, $g^{(k)}(x)$ is the element $g(g(g(\cdots g(x))))$, where the function g is applied k times. If there exists $m \geq 1$ for which $g^{(m)}(x) = x$, then the orbit of x is $\{x, g(x), \ldots, g^{m-1}(x)\}$. If $g(x) = x$, then x is called a *fixed point* of g and has orbit $\{x\}$.

Lemma 14 *Let g be a function from S to S and let x be an element of S. Suppose for some integer $n \geq 1$, $g^{(n)}(x) = x$ and let m be the smallest positive integer for which $g^{(m)}(x) = x$. Then m divides n.*

Proof. Let $n = qm + r$ where $0 \leq r < m$. Then

$$\begin{aligned} x = g^{(n)}(x) &= g^{(r+qm)}(x) \\ &= g^{(r+m+m+\cdots+m)}(x) \\ &= g^{(r)}(g^{(m)}(g^{(m)}(\cdots g^{(m)}(x)\cdots))) \\ &= g^{(r)}(x), \end{aligned}$$

and since $0 \leq r < m$, we must have $r = 0$ by the minimality of m. Hence m divides n.
◇

Corollary 15 *Let S be a finite set and g be a function from S to S. Suppose n is an integer such that $g^{(n)}(x) = x$ for all x in S. Then the size of every orbit divides n.*

Notice that under the conditions described in the corollary above, the orbits partition S into disjoint subsets. Hence the size of S is the sum of the sizes of the orbits. When n is prime, the situation is especially simple.

Corollary 16 *Let S be a finite set and suppose there exists a prime number p for which $g^{(p)}(x) = x$ for all x in S. Then every orbit either has size 1 or size p. Consequently, if F is the set of all fixed points of g, then*

$$|S| \equiv |F| \pmod{p}.$$

As our first application, we have

Proof 2 of Theorem 13. Let S be the set of k-element subsets of $\{1, \ldots, p\}$. For X in S, $X = \{x_1, x_2, \ldots, x_k\}$, define $g(X) = \{x_1 + 1, x_2 + 1, \ldots, x_k + 1\}$, where all sums are reduced modulo p. Now since $g^{(p)}(X) = X$ for all X, and S has no fixed points of g (why?), then by Corollary 16, we have $|S| \equiv 0 \pmod{p}$.
◇

The next theorem is one of the most useful in number theory.

Theorem 17 (Fermat's Little Theorem) *If p is prime, then for any integer a, p divides $a^p - a$.*

Proof 1. The classical proof uses the Binomial Theorem (Identity 133), Theorem 13 and proceeds by induction on a. If p divides $a^p - a$ (as it clearly does when $a = 0$ or $a = 1$) then it must divide $(a+1)^p - (a+1)$ since $(a+1)^p = \sum_{k=0}^{p}\binom{p}{k}a^k = 1 + a^p + \sum_{k=1}^{p-1}\binom{p}{k}a^k \equiv 1 + a^p + 0 \equiv a + 1 \pmod{p}$.
◇

Proof 2. This proof is more combinatorial in spirit.

> **Question:** How many ways can the numbers $\{1, \ldots, p\}$ each be assigned one of a colors, where not all numbers are allowed to be assigned the same color.

> **Answer 1:** There are $a^p - a$ such colorings since all a monochromatic colorings must be removed from consideration.

elements	1	2	3	4	5
original coloring	B	R	R	Y	B
equivalent coloring	B	B	R	R	Y
equivalent coloring	Y	B	B	R	R
equivalent coloring	R	Y	B	B	R
equivalent coloring	R	R	Y	B	B

Figure 8.2. If the numbers $1, 2, 3, 4, 5$ are assigned the colors *Blue, Red, Red, Yellow,* and *Blue* respectively, equivalent colorings are obtained by cyclically shifting the assignment.

Answer 2: For each such coloring and for $0 \le j < p$, consider the equivalence class of colorings obtained by shifting all the colors j positions to the right modulo p. For example when $p = 5$, the coloring $(Blue, Red, Red, Yellow, Blue)$ generates the class of colorings in Figure 8.2. By the second proof of Theorem 13, all of the colorings in a class must be distinct. (For any utilized color, the subset of numbers receiving that color must be different in each coloring.) Hence the set of colorings can be partitioned into classes, all of which have size p. So the total number of colorings is p times the number of equivalence classes.

Since p is a factor in Answer 2, it must also divide Answer 1. Hence p divides $a^p - a$. $\diamond$

Euler proved a more general identity. For a positive integer n, the *totient* function $\phi(n)$ counts the number of integers in $\{1, \ldots, n\}$ that are relatively prime (i.e., have no common prime factors) with n. Euler's Theorem states that if a and n are relatively prime, then

$$a^{\phi(n)} \equiv 1 \pmod{n}.$$

Although we do not know of a combinatorial proof of this theorem, we would love to see one! Nevertheless, we can prove other facts about $\phi(n)$ combinatorially.

Theorem 18 *If $n = p_1^{e_1} p_2^{e_2} \cdots p_t^{e_t}$, where the p_is are distinct primes and all exponents are positive integers, then*

$$\phi(n) = n \left(1 - \frac{1}{p_1} \right) \left(1 - \frac{1}{p_2} \right) \cdots \left(1 - \frac{1}{p_t} \right).$$

This theorem makes sense intuitively since among the n numbers $1, 2, \ldots, n$, $1/p_1$th of them have p_1 as a prime factor. The $n(1 - \frac{1}{p_1})$ term is the size of the set after removing all multiples of p_1. Intuitively, one would expect that $1/p_2$th of these remaining numbers should be divisible by p_2, so after eliminating these, the remaining set should have size

$$n \left(1 - \frac{1}{p_1} \right) \left(1 - \frac{1}{p_2} \right).$$

Continuing in this manner, the subset of these numbers not divisible by any of the prime factors of n should be

$$n \left(1 - \frac{1}{p_1} \right) \left(1 - \frac{1}{p_2} \right) \cdots \left(1 - \frac{1}{p_t} \right).$$

A more rigorous combinatorial proof uses the principle of inclusion-exclusion.

Question: How many numbers in $\{1, \ldots, n\}$ are relatively prime to n?

Answer 1: By definition, $\phi(n)$.

Answer 2: We use inclusion-exclusion here. From the set $\{1, \ldots, n\}$, we first throw away the n/p_1 multiples of p_1 and the n/p_2 multiples of $p_2, \ldots$, and the n/p_t multiples of p_t. Then we add back the n/p_1p_2 numbers that are divisible by p_1 and p_2, the n/p_1p_3 numbers that are divisible by p_1 and p_3, and so on. Then we subtract off the $n/p_1p_2p_3$ numbers that are divisible by p_1, p_2, and p_3, and so on. When the dust settles, our subset of $\{1, \ldots, n\}$ with none of these prime factors has size

$$n - \left(\frac{n}{p_1} + \frac{n}{p_2} + \frac{n}{p_3} + \cdots \right) + \left(\frac{n}{p_1 p_2} + \frac{n}{p_1 p_3} + \cdots \right)$$
$$- \left(\frac{n}{p_1 p_2 p_3} + \cdots \right) + \cdots + (-1)^t \frac{n}{p_1 p_2 \cdots p_t}$$

which simplifies to

$$n \left(1 - \frac{1}{p_1} \right) \left(1 - \frac{1}{p_2} \right) \cdots \left(1 - \frac{1}{p_t} \right).$$

We leave as an exercise to prove

Corollary 19 *If x and y are integers with no common prime factors, then $\phi(xy) = \phi(x)\phi(y)$.*

The next identity sums $\phi(d)$ over all positive divisors of n.

Identity 219 $\displaystyle\sum_{d|n} \phi(d) = n.$

Question: How many numbers are in the set $\{\frac{1}{n}, \frac{2}{n}, \ldots, \frac{n}{n}\}$?

Answer 1: Obviously, n.

Answer 2: Each such fraction can be put in lowest terms of the form $\frac{c}{d}$ where d is a divisor of n, and c is relatively prime to d. For each denominator d there are $\phi(d)$ relatively prime numerators. Altogether the number of fractions is $\sum_{d|n} \phi(d)$.

Here is another fundamental theorem from number theory.

Identity 220 (Wilson's Theorem) *If p is prime, then*

$$(p-1)! \equiv p - 1 \pmod{p}.$$

Question: How many permutations of $\{0, 1, \ldots, p-1\}$ have exactly one cycle?

Answer 1: $(p-1)!$

Answer 2: Let S be the set of permutations of $\{0, 1, \ldots, p-1\}$ with exactly one cycle, and define the function g on S as follows. For any permutation $\pi = (0, a_1, a_2, \ldots, a_{p-1})$ in S, define $g(\pi) = (1, a_1 + 1, a_2 + 1, \ldots, a_{p-1} + 1)$, where all sums are reduced modulo p. Clearly $g(\pi)$ is in S (although it will not be in standard form since it does not begin with 0) and $g^{(p)}(\pi) = \pi$. Thus by Corollary 16, $|S|$ is congruent mod p to the number of fixed points of g. It remains to show that S has precisely $p-1$ fixed points of g, namely those "arithmetic progression" permutations $(0, a, 2a, 3a, \ldots, (p-1)a)$, where $1 \leq a \leq p-1$ and all terms are reduced mod p. Such permutations are fixed points of g (since adding one to each entry does not spoil the arithmetic progression property). Conversely, if $\pi = (0, a_1, a_2, \ldots, a_{p-1})$ is a fixed point of g, then

$$(0, a_1, a_2, a_3, \ldots, a_{p-1}) = \pi = g^{(a_1)}(\pi) = (a_1, 2a_1, a_2 + a_1, \ldots, a_{p-1} + a_1).$$

On the left side, $\pi(a_1) = a_2$. On the right side, $\pi(a_1) = 2a_1$. So $a_2 = 2a_1$. Continuing on the left side, $\pi(a_2) = a_3$ and on the right side $\pi(a_2) = a_2 + a_1 = 3a_1$. So $a_3 = 3a_1$. In general $a_k = ka_1$. Hence g has $p-1$ fixed points, as desired.

Lagrange's Theorem may be the most important theorem concerning finite groups. It states that if a group G with n elements has a subgroup H with d elements, then d must be a divisor of n. The converse of this theorem is false. That is, if d divides n, there does not necessarily exist a subgroup of G of size d. However, when d is prime, the converse holds.

Theorem 20 *If G is a group of order n and the prime p divides n, then G has a subgroup of order p.*

Proof. Suppose that G has n elements, where n is divisible by the prime p. Define

$$\mathcal{S} = \{(x_1, x_2, \ldots, x_p) | x_i \in G, x_1 x_2 \cdots x_p = e\},$$

where e denotes the identity element of G.

Question: How many elements does $\mathcal{S}$ have?

Answer 1: There are n^{p-1} elements in $\mathcal{S}$ since we can arbitrarily choose the terms $x_1, x_2, \ldots, x_{p-1}$ (n choices each), which then forces $x_p = (x_1 x_2 \cdots x_{p-1})^{-1}$.

Answer 2: Notice that if $s_1 = (x_1, x_2, \ldots, x_p) \in \mathcal{S}$, then we have $x_2 x_3 \cdots x_p = x_1^{-1}$ which implies $x_2 x_2 \cdots x_p x_1 = e$. It follows that $s_2 = (x_2, x_3, \ldots, x_p, x_1)$ is also in $\mathcal{S}$. Continuing in this manner, $s_3 = (x_3, x_4, \ldots, x_p, x_1, x_2)$, $s_4 = (x_4, x_5, \ldots, x_p, x_1, x_2, x_3), \ldots, s_p = (x_p, x_1, \ldots, x_{p-1})$ also belong to $\mathcal{S}$. Since p is prime, the elements $s_1, s_2, \ldots, s_p$ must be distinct unless $x_1 = x_2 = \cdots = x_p$. Consequently $|\mathcal{S}| = |\mathcal{S}_1| + |\mathcal{S}_2|$ where $\mathcal{S}_1$ is the set of elements of $\mathcal{S}$ of the form $(x, x, \ldots, x)$ and $\mathcal{S}_2$ consists of all other elements of $\mathcal{S}$. The size of $\mathcal{S}_2$ is a multiple of p since its elements can be partitioned into cyclic shifts of length p. Thus $|\mathcal{S}_1| = n^{p-1} - pk$ where k is some integer. Since p divides n, we conclude that $|\mathcal{S}_1|$ is a multiple of p. Furthermore, the size of $\mathcal{S}_1$ is nonzero since $(e, e, \ldots, e)$ is a member. Thus G has at least $p-1$ nonidentity elements x for which $x^p = e$. For any such x, the set $H = \{e, x, x^2, \ldots, x^{p-1}\}$ is a cyclic subgroup of G with p elements.   ◇

8.3 GCDs Revisited

In Chapter 1, we proved that the greatest common divisor of the traditional Fibonacci numbers ($F_0 = 0$, $F_1 = 1$, $F_n = F_{n-1} + F_{n-2}$) satisfy the identity

$$\gcd(F_n, F_m) = F_{\gcd(n,m)}.$$

We show that this phenomenon remains true for other Lucas sequences of the first kind.

Theorem 21 *Let s, t be nonnegative relatively prime integers and consider the sequence $U_0 = 0, U_1 = 1$, and for $n \geq 2$, $U_n = sU_{n-1} + tU_{n-2}$. Then*

$$\gcd(U_n, U_m) = U_{\gcd(n,m)}.$$

For convenience, we define for $n \geq 0$,

$$u_n = U_{n+1},$$

which by Combinatorial Theorem 4 counts colored tilings of length n with s colors for squares and t colors for dominoes. Fortuitously, in Chapter 3 we already derived the identities needed to prove the following lemma.

Lemma 22 *For all $m \geq 1$, U_m and tU_{m-1} are relatively prime.*

Proof. First we claim that U_m is relatively prime to t. This is easy to see algebraically since for all $m \geq 1$, $U_m = sU_{m-1} + tU_{m-2} \equiv sU_{m-1} \bmod t$. Thus by Euclid's algorithm discussed on page 11, $\gcd(U_m, t) = \gcd(t, s^{m-1}) = 1$. For the combinatorial purist, we can condition on the location of the last colored domino (if any exist). Identity 75 says (after letting $c = s$ and re-indexing),

$$U_m = s^{m-1} + t \sum_{j=1}^{m-2} s^{j-1} U_{m-1-j}.$$

Consequently, if $d > 1$ is a divisor of U_m and t, then d must also divide s^{m-1}, which is impossible since s and t are relatively prime.

Next we claim that U_m and U_{m-1} are relatively prime. This follows from Identity 87 since if $d > 1$ divides U_m and U_{m-1}, then d must divide t^{m-1}. But this is impossible since U_m and t are relatively prime.

Thus since $\gcd(U_m, t) = 1$ and $\gcd(U_m, U_{m-1}) = 1$, then $\gcd(U_m, tU_{m-1}) = 1$, as desired. $\diamond$

To prove Theorem 21, we will need one more identity, inspired by Euclid's algorithm for computing greatest common divisors. The next identity may look formidable at first, but it makes sense when viewed combinatorially.

Identity 221 *If $n = qm + r$, where $0 \leq r < m$, then*

$$U_n = (tU_{m-1})^q U_r + U_m \sum_{j=1}^{q} (tU_{m-1})^{j-1} U_{(q-j)m+r+1}.$$

Question: How many colored $(qm + r - 1)$-tilings exist?

Answer 1: $u_{qm+r-1} = U_{qm+r} = U_n$.

Answer 2: First we count all such colored tilings that are unbreakable at every cell of the form $jm - 1$, where $1 \leq j \leq q$. Such a tiling must have a colored domino starting on cell $m - 1, 2m - 1, \ldots, qm - 1$; these dominoes can be chosen t^q ways. Before each of these dominoes is an arbitrary $(m - 2)$-tiling that can each be chosen u_{m-2} ways. Finally, cells $qm + 1, \ldots, qm + r - 1$ can be tiled u_{r-1} ways. See Figure 8.3. Consequently, the number of colored tilings with no $jm - 1$ breaks is $t^q(u_{m-2})^q u_{r-1} = (tU_{m-1})^q U_r$.

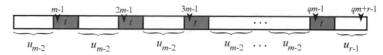

Figure 8.3. There are $(tU_{m-1})^q U_r$ colored $(qm + r - 1)$-tilings with no breaks at any cells of the form $jm - 1$ where $1 \leq j \leq q$.

Next, we partition the remaining colored tilings according to the first breakable cell of the form $jm - 1$, $1 \leq j \leq q$. By similar reasoning as before, this can be done $(tU_{m-1})^{j-1} U_m U_{(q-j)m+r+1}$ ways. See Figure 8.4. Altogether, the number of colored tilings is $(tU_{m-1})^q U_r + U_m \sum_{j=1}^{q} (tU_{m-1})^{j-1} U_{(q-j)m+r+1}$.

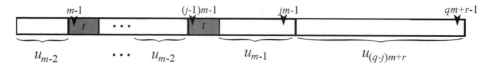

Figure 8.4. There are $(tU_{m-1})^{j-1} U_m U_{(q-j)m+r+1}$ colored $(qm+r-1)$-tilings that are breakable at cell $jm - 1$, but not at cells of the form $im - 1$ where $1 \leq i < j$.

The previous identity explicitly shows that U_n is an integer combination of U_m and U_r. Consequently, d is a common divisor of U_n and U_m if and only if d divides U_m and $(tU_{m-1})^q U_r$. But by Lemma 22, since U_m is relatively prime to tU_{m-1}, d must be a common divisor of U_m and U_r. Thus U_n and U_m have the same common divisors (and hence the same gcd) as U_m and U_r. In other words,

Corollary 23 *If $n = qm + r$, where $0 \leq r < m$, then*

$$\gcd(U_n, U_m) = \gcd(U_m, U_r).$$

But wait!! This corollary is the same as Euclid's algorithm, with Us inserted everywhere. This proves Theorem 21 by following the same steps as Euclid's algorithm. The $\gcd(U_n, U_m)$ will eventually reduce to $\gcd(U_g, U_0) = \gcd(U_g, 0) = U_g$, where g is the greatest common divisor of m and n.

8.4   Lucas' Theorem

In Section 5.5, we found an ingenious method for determining the parity of $\binom{n}{k}$ and proved that for a fixed n, the number of odd values of $\binom{n}{k}$ is 2 raised to the number of ones in the binary expansion of n. For instance, $82 = 64 + 16 + 2 = (1010010)_2$ so there are 2^3 odd values of $\binom{82}{k}$. Here we present another proof of this fact and the amazing generalization modulo an arbitrary prime due to Edouard Lucas of Lucas number fame.

We begin with the following generalization of Theorem 13 to prime powers.

Theorem 24 *Let p be prime. For any $\alpha \geq 1$ and any $0 < k < p^\alpha$,*

$$\binom{p^\alpha}{k} \equiv 0 \pmod{p}.$$

Proof. Again by Identity 130 of Chapter 5, we know that

$$k \binom{p^\alpha}{k} = p^\alpha \binom{p^\alpha - 1}{k - 1}.$$

Since $0 < k < p^\alpha$, the largest power of p that divides k is at most $p^{\alpha-1}$. Consequently, p divides $\binom{p^\alpha}{k}$. ◇

Recall that polynomials with integer coefficients

$$f(x) = \sum_{n \geq 0} a_n x^n \quad \text{and} \quad g(x) = \sum_{n \geq 0} b_n x^n$$

are congruent modulo p if their coefficients are congruent term by term. Specifically, $f(x) \equiv g(x) \pmod{p}$ if for all $n \geq 0$, $a_n \equiv b_n \pmod{p}$. For example,

$$x^4 + 4x^3 + 6x^2 + 4x + 1 \equiv x^4 + 1 \pmod{2}.$$

The next lemma is sometimes facetiously referred to as the "freshman's binomial theorem."

Lemma 25 *For p prime and $\alpha \geq 0$,*

$$(1 + x)^{p^\alpha} \equiv 1 + x^{p^\alpha} \pmod{p}.$$

Proof. As in the proof of Theorem 17, the (legitimate) binomial theorem and modular arithmetic give

$$(1 + x)^{p^\alpha} = \sum_{k=0}^{p^\alpha} \binom{p^\alpha}{k} x^k \equiv 1 + x^{p^\alpha} \pmod{p},$$

since all terms vanish modulo p except when $k = 0$ and $k = p^\alpha$. ◇

To find the parity of $\binom{82}{k}$, write $82 = 64 + 16 + 2$ and apply Lemma 25 with $p = 2$.

$$\sum_{k=0}^{82} \binom{82}{k} x^k = (1 + x)^{82}$$
$$= (1 + x)^{64}(1 + x)^{16}(1 + x)^2$$
$$\equiv (1 + x^{64})(1 + x^{16})(1 + x^2) \pmod{2}$$
$$\equiv 1 + x^2 + x^{16} + x^{18} + x^{64} + x^{66} + x^{80} + x^{82} \pmod{2}.$$

Thus $\binom{82}{k}$ has the same parity as the coefficient of x^k in the last expression. For instance, $\binom{82}{18} \equiv 1 \pmod{2}$ is odd, whereas $\binom{82}{20} \equiv 0 \pmod{2}$ is even. Values of $\binom{82}{k}$ for which $\binom{82}{k}$ is odd are precisely those k that can be expressed as $64a + 16b + 2c$ where $a, b, c \in \{0, 1\}$. Hence there are $2^3 = 8$ odd values of $\binom{82}{k}$. Generalizing this example, if the binary expansion of n is $\sum_{i=0}^{t} b_i 2^i$ where $b_i = 0$ or 1, then the number of odd values of $\binom{n}{k}$ is

$$\prod_{i=0}^{t}(1 + b_i) = 2^{\text{number of 1s in the binary expansion of } n}.$$

We are now ready to state and prove Lucas' Theorem.

Theorem 26 (Lucas' Theorem) *For any prime p, we can determine $\binom{n}{k}$ (mod p) from the base p expansions of n and k. Specifically, if $n = \sum_{i=0}^{t} b_i p^i$ and $k = \sum_{i=0}^{t} c_i p^i$ where $0 \le b_i, c_i < p$, then*

$$\binom{n}{k} \equiv \prod_{i=0}^{t} \binom{b_i}{c_i} \pmod{p}.$$

Consider the situation where $n = 97$, $k = 35$, and $p = 5$,

$$97 = 3 \cdot 5^2 + 4 \cdot 5 + 2 = (3\ 4\ 2)_5,$$

$$35 = 1 \cdot 5^2 + 2 \cdot 5 + 0 = (1\ 2\ 0)_5.$$

Consequently, Lucas' Theorem implies

$$\binom{97}{35} \equiv \binom{3}{1}\binom{4}{2}\binom{2}{0} = 18 \equiv 3 \pmod{5}.$$

Whereas, since $38 = (1\ 2\ 3)_5$,

$$\binom{97}{38} \equiv \binom{3}{1}\binom{4}{2}\binom{2}{3} = 0 \pmod{5}.$$

Lucas' Theorem implies that when $n = \sum_{i=0}^{t} b_i p^i$ and $k = \sum_{i=0}^{t} c_i p^i$, $\binom{n}{k}$ is a multiple of p if and only if $c_i > b_i$ for some $0 \le i \le t$. Thus for $\binom{n}{k}$ to avoid being a multiple of p, we must have for each i, $c_i \in \{0, 1, \ldots, b_i\}$. Consequently, $\binom{n}{k}$ is *not* a multiple of p for exactly $\prod_{i=0}^{t}(1 + b_i)$ values of k.

We illustrate the proof of Lucas' Theorem with an example. When $n = 97$ and $p = 5$, we determine $\binom{97}{k}$ (mod 5) as follows. By the binomial theorem, $\binom{97}{k}$ is the x^k coefficient of $(1+x)^{97}$. Lemma 25 with $p = 5$ gives

$$(1+x)^{97} = (1+x)^{3 \cdot 25}(1+x)^{4 \cdot 5}(1+x)^{2 \cdot 1}$$
$$= (1+x)^{25}(1+x)^{25}(1+x)^{25}(1+x)^{5}(1+x)^{5}$$
$$\times (1+x)^{5}(1+x)^{5}(1+x)(1+x)$$
$$\equiv (1+x^{25})(1+x^{25})(1+x^{25})(1+x^{5})(1+x^{5})$$
$$\times (1+x^{5})(1+x^{5})(1+x^{1})(1+x^{1}) \pmod{p}.$$

Hence $\binom{97}{k}$ will be congruent to the x^k coefficient of the last expression modulo 5. For instance, the x^{35} term in the last expression is the number of ways we can reach a total of 35 cents with three distinct quarters, four distinct nickels, and two distinct pennies at our disposal. Since a total of 35 cents requires exactly one 25-cent coin, two 5-cent coins and zero 1-cent coins, there are $\binom{3}{1}\binom{4}{2}\binom{2}{0}$ coin combinations. Thus $\binom{97}{35} \equiv \binom{3}{1}\binom{4}{2}\binom{2}{0} \pmod{5}$, as desired. By the same reasoning, there are no ways to achieve 38 cents using these coins since there are not enough pennies. Consequently, we have $\binom{97}{38} \equiv 0 \pmod{5}$.

Another strategy to prove Lucas' Theorem that generalizes the approach taken in Theorem 26 is outlined in the exercises.

8.5 Notes

The second proof of Fermat's Little Theorem is attributed to J. Peterson in Dickson's *History of the Theory of Numbers* [23]. The partial converse to Lagrange's Theorem is due to Cauchy; the combinatorial proof we provide is due to McKay [35].

We have only scratched the surface of combinatorial congruences. For further information see Chapter 1 of Stanley [51] and papers by Rota and Sagan [47], Gessel [26], Sagan [49], and Smith [50]. See also Erdős and Graham [24] and the survey article by Pomerance and Sárközy [42].

8.6 Exercises

Prove each of the identities and theorems below by a direct combinatorial argument.

Identity 222 *For $n \geq 0$, $\sum_{k=1}^{n} k^4 = \binom{n+1}{2} + 14\binom{n+1}{3} + 36\binom{n+1}{4} + 24\binom{n+1}{5}$.*

Identity 223 *For $n \geq 1$, $\sum_{k=1}^{n-1} k^2 = \frac{1}{4}\binom{2n}{3}$.*

Identity 224 *For $0 \leq r, s \leq 1$ and $n \geq 0$, $\binom{2n+r}{2k+s} \equiv \binom{n}{k}\binom{r}{s} \pmod{2}$.*

To prove Identity 224, count palindromic binary strings depending on the length of the string and the parity of the number of 1s as indicated in the four cases below.

1. $\binom{2n}{2k} \equiv \binom{n}{k} \pmod{2}$.

2. $\binom{2n+1}{2k+1} \equiv \binom{n}{k} \pmod{2}$.

3. $\binom{2n+1}{2k} \equiv \binom{n}{k} \pmod{2}$.

4. $\binom{2n}{2k+1} \equiv 0 \pmod{2}$.

Identity 225 *For $n, k \geq 0$ and p prime, $\binom{pn}{pk} \equiv \binom{n}{k} \pmod{p}$.*

Identity 226 *For $0 \leq k \leq n$, $0 \leq s \leq r$, and p prime, $\binom{pn+r}{pk+s} \equiv \binom{n}{k}\binom{r}{s} \pmod{p}$.*

Deduce Lucas' Theorem from Identities 225 and 226.

Identity 227 *For $0 \leq k \leq n$ and p prime, $\binom{pn}{pk} \equiv \binom{n}{k} \pmod{p^2}$.*

Identity 228 *For p prime, the pth Lucas number satisfies*

$$L_p \equiv 1 \pmod{p}.$$

Identity 229 *For p prime, $L_{2p} \equiv 3 \pmod{p}$.*

Identity 230 *For distinct primes p and q, $L_{pq} \equiv 1 + (L_q - 1)q \pmod{p}$.*

Theorem 27 *If m divides n, then U_m divides U_n.*

Theorem 28 *L_m divides F_{2km}.*

Theorem 29 *L_m divides $L_{(2k+1)m}$.*

Uncounted Identities

The identities listed below are in need of combinatorial proof.

1. For $p \geq 5$ and p prime, $\binom{pn}{pk} \equiv \binom{n}{k} \pmod{p^3}$.

2. If m, n are odd integers, with $\gcd(m, n) = g$, then $\gcd(L_m, L_n) = L_g$.

   More generally, suppose $m = 2^a m'$ and $n = 2^b n'$, and $\gcd(m, n) = g$, with m', n' odd and $a, b \geq 0$. Prove

3. $\gcd(L_m, L_n) = \begin{cases} L_g & \text{if } a = b, \\ 2 & \text{if } a \neq b \text{ and } 3|g, \\ 1 & \text{otherwise.} \end{cases}$

4. $\gcd(F_m, L_n) = \begin{cases} L_g & \text{if } a > b, \\ 2 & \text{if } a \leq b \text{ and } 3|g, \\ 1 & \text{otherwise.} \end{cases}$

CHAPTER 9

Advanced Fibonacci & Lucas Identities

9.1 More Fibonacci and Lucas Identities

We end this book as we began it, by exploring more Fibonacci and Lucas identities (Lucas might say that we've come "full circle"!) We include some of the proofs that we found particularly challenging. As a climax, we add a dose of probability to obtain combinatorial proofs of the *Binet formulas*

$$F_n = \frac{1}{\sqrt{5}} \left[\left(\frac{1 + \sqrt{5}}{2} \right)^n - \left(\frac{1 - \sqrt{5}}{2} \right)^n \right]$$

and

$$L_n = \left(\frac{1 + \sqrt{5}}{2} \right)^n + \left(\frac{1 - \sqrt{5}}{2} \right)^n.$$

We conclude with some known identities that, as far as we know, have not yet succumbed to combinatorial interpretation.

Recall by Combinatorial Theorem 1 that f_n is the number of square and domino tilings of a length n board. The first identity is a warm-up to remind us how these Fibonacci tilings function.

Identity 231 *For $n, m \geq 2$,*

$$f_n f_m - f_{n-2} f_{m-2} = f_{n+m-1}.$$

Set 1: The set of ordered pairs (A, B), where A is an n-tiling, B is an m-tiling and either A or B must end with a square. Discarding the tiling pairs where both end in a domino gives us $f_n f_m - f_{n-2} f_{m-2}$ such tilings.

Set 2: The set of $(n + m - 1)$-tilings. This set has size f_{n+m-1}.

Correspondence: We consider two cases, as illustrated in Figure 9.1. If A ends with a square, then we append B to A after removing the last square of A. This creates an $(n + m - 1)$-tiling that is breakable at cell $n - 1$. Otherwise, A must end with a domino and B must end with a square. Here we append B to A with the last square of B removed. This creates an $(n + m - 1)$-tiling that is unbreakable at cell $n - 1$.

A ends in a square:

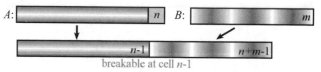

breakable at cell n-1

A ends in a domino and *B* ends in a square:

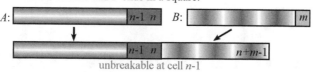

unbreakable at cell n-1

Figure 9.1. An n-tiling, m-tiling pair where at least one tiling ends with a square corresponds to a single $(n + m - 1)$-tiling.

The next identity is more complicated, but follows a similar logic.

Identity 232

$$f_{n-1}^3 + f_n^3 - f_{n-2}^3 = f_{3n-1}.$$

Set 1: The set of ordered triples (A, B, C), where A, B, and C are $(n-1)$-tilings or A, B, and C are n-tilings with at least one ending with a square. Discarding the triples of n-tilings that all end in a domino shows that this set has size $f_{n-1}^3 + f_n^3 - f_{n-2}^3$.

Set 2: The set of $(3n - 1)$-tilings. There are f_{3n-1} such tilings.

Correspondence: This time we need four cases, as illustrated in Figure 9.2. For the first three cases, we assume that (A, B, C) is a triple of n-tilings. If A ends in a square, then we append B then C to A after removing the last square of A. This creates a $(3n - 1)$-tiling that is breakable at cells $n - 1$ and $2n - 1$. If A ends in a domino and B ends in a square, then we append B then C to A after removing the last square of B. This creates a $(3n - 1)$-tiling that is unbreakable at cell $n - 1$ and breakable at cell $2n - 1$. If A and B end with a domino and C ends with a square, then we do the same thing as before, but remove the last square of C. This creates a $(3n - 1)$-tiling that is unbreakable at cells $n - 1$ and $2n - 1$. For the fourth case, where (A, B, C) is a triple of $(n - 1)$-tilings, we need to generate $(3n - 1)$-tilings that are breakable at cell $n - 1$ and unbreakable at cell $(2n - 1)$. We do this by appending B then C to A with an extra domino inserted between B and C.

As a warmup for the more difficult identity that follows, we first present this simpler one to illustrate the main idea.

Identity 233

$$f_n^2 + 2(f_0^2 + f_1^2 + \cdots + f_{n-1}^2) = f_{2n+1}.$$

Question: How many ways can a $(2n+1)$-board be tiled using squares and dominoes?

Answer 1: f_{2n+1}.

Answer 2: Condition on the location of the "middlest" square, i.e., the square that is closest to the center of the tiling at cell $n + 1$. Since $2n + 1$ is odd, at least one middlest square must exist. In fact, the middlest square is unique since if we have

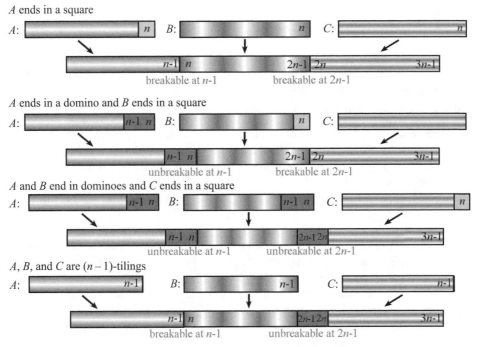

Figure 9.2. A triple-tiling identity for $f_{n-1}^3 + f_n^3 - f_{n-2}^3 = f_{3n-1}$.

squares at cells $n+1-k$ and $n+1+k$ for $k \geq 1$, then a closer square would have to exist among the $2k-1$ cells between them. There are f_n^2 tilings with the middlest square at cell $n+1$. See Figure 9.3. For the middlest square to occur at cell j for $1 \leq j \leq n$, cells $j+1$ through $2n+2-j$ must be covered with dominoes. The remaining cells can be tiled f_{j-1}^2 ways. Similarly, for $1 \leq j \leq n$, there are f_{j-1}^2 tilings with middlest square at cell $2n+2-j$. Altogether, we have $f_n^2 + 2\sum_{j=1}^n f_{j-1}^2$ tilings.

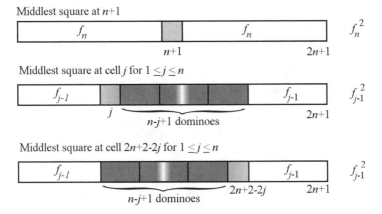

Figure 9.3. Conditioning on the "middlest" square to show that $f_n^2 + 2(f_0^2 + f_1^2 + \cdots + f_{n-1}^2) = f_{2n+1}$.

Identity 234

$$5[f_0^2 + f_2^2 + \cdots + f_{2n-2}^2] = f_{4n-1} + 2n.$$

Set 1: Let $\mathcal{S}$ denote the set of tiling pairs (A, B) where A and B both have length $2j$ for some $0 \leq j \leq n - 1$. Clearly $|\mathcal{S}| = \sum_{j=0}^{n-1} f_{2j}^2$.

Set 2: Let $\mathcal{T}$ denote the set of $(4n - 1)$-tilings. $|\mathcal{T}| = f_{4n-1}$.

Correspondence: We create an almost 1-to-5 correspondence between $\mathcal{S}$ and $\mathcal{T}$. Specifically, for a tiling pair (A, B) of $\mathcal{S}$ except those consisting entirely of j dominoes for $0 \leq j \leq n - 1$, we identify five tilings of $\mathcal{T}$. In the n exceptional cases, we identify only three tilings of $\mathcal{T}$ which accounts for the discrepancy of $2n$. Our correspondence will be reversible according to the "middlest square" defined in the previous proof.

For every tiling pair (A, B) in $\mathcal{S}$, we first generate tilings T_1, T_2, and T_3 in $\mathcal{T}$ by inserting a square s and a sequence of k dominoes, d^k, as described below and indicated in Figure 9.4. $T_1 = Asd^{2n-2j-1}B$, i.e., T_1 is the tiling that begins with A followed by a single square followed by $2n - 2j - 1$ dominoes followed by tiling B. The "middle section" of T_1 consists of the middlest square followed by an odd number of dominoes. Using the same notation, $T_2 = Ad^{2n-2j-1}sB$ whose middle section has an odd number of dominoes and ends with the middlest square. We pause to note that T_1 and T_2 are all of the tilings of $\mathcal{T}$ with an odd number of dominoes in the middle section (or equivalently, an even number of cells in the "left section" and "right section"). Next $T_3 = sAsd^{2n-2j-2}sB$ whose middle section consists of the middlest square followed by an even number of dominoes and whose left and right sections (which necessarily have the same odd length) both begin with squares.

Tilings T_4 and T_5 are created by tiling pairs (A, B) from $\mathcal{S}$ that are not all dominoes. From the tail swapping technique of Chapter 1 introduced on page 7, we know that if A and B are both $(2j)$-tilings for some $j \geq 1$, (not consisting of all dominoes) then we can associate the tiling pair (A', B') where A' is a $(2j - 1)$-tiling and B' is a $(2j + 1)$-tiling. Define $T_4 = dA'sd^{2n-2j-2}B'$, where the left section begins with a domino, and the middle section consists of the middlest square followed by an even

Create three $(4n - 1)$ tilings from every pair of $2j$-tilings (A, B)

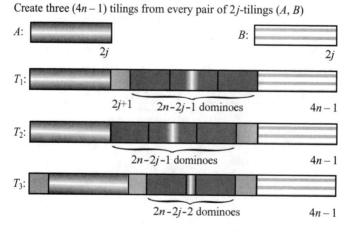

Figure 9.4. Tilings T_1, T_2, T_3.

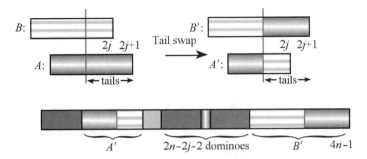

Figure 9.5. Tiling T_4 can only be created in pairs (A,B) containing at least one square.

number of dominoes. See Figure 9.5. It remains to construct those $(4n-1)$-tilings of the following two forms:

— left section begins with a square, middlest section begins with the middlest square, followed by an even number of dominoes, and the right section begins with a domino,

and

— middle section ends with the middlest square, preceded by an even number of dominoes.

T_5, illustrated in Figure 9.6, will be of one of these two types, depending on whether B begins with a square or domino. Specifically, if B begins with a square, so that $B = sB^*$, then $T_5 = sAsd^{2n-2j-2}dB^*$. Notice that both the left section sA and the right section dB^* of T_5 both have $2j+1$ cells, an odd number. If B begins with a domino, so that $B = dB^{**}$, then since A and B are not all dominoes, we can tail swap A with B^{**} to create A' and B' which are both $2j-1$ tilings. Thus $T_5 = A'd^{2n-2j}sB'$ has the desired form.

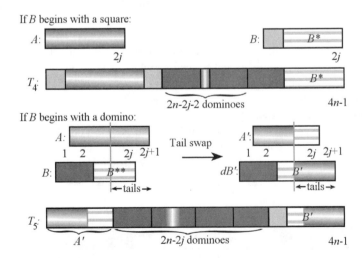

Figure 9.6. Tiling T_5 can only be created in pairs (A, B) containing at least one square. The resulting $(4n-1)$-tiling depends on whether B begins with a square or a domino.

9.2   Colorful Identities.

The combinatorial proofs of the identities in this section may be the most challenging ones in the book. For these identities we will need a combinatorial interpretation of $2^n f_n$. There are $2^n f_n$ ways to mark some of the cells of an n-board with an "X", then tile the board with "transparent" squares or dominoes. Looking through the tiles, we see that we have two types of squares, which we call *black* or *white*, and four types of dominoes, which we call *red, yellow, green,* or *violet*. See Figure 9.7.

white   black   red   yellow   green   violet

Figure 9.7. The six colored tiles that arise from marking some of the cells of an n-board with an X and then tiling the board with transparent squares and dominoes.

Identity 235 $\sum_{t=0}^{n} \binom{n}{t} 5^{\lfloor \frac{t}{2} \rfloor} = 2^n f_n.$

> **Question:** In how many ways can we tile an n-board with two types of colored squares and four types of colored dominoes?
>
> **Answer 1:** By the previous paragraph, there are $2^n f_n$ such tilings.
>
> **Answer 2:** For every t-element subset of $\{1, 2, \dots, n\}$, we generate $5^{\lfloor \frac{t}{2} \rfloor}$ different colored tilings in the manner described below.
>
> Let $a_1 < a_2 < \cdots < a_t$ be a subset of $\{1, 2, \dots, n\}$, and suppose that t is even. This gives rise to $t/2$ disjoint intervals $I_1 = [a_1, a_2]$, $I_2 = [a_3, a_4], \dots$, $I_{t/2} = [a_{t-1}, a_t]$. Any cell not belonging to one of these intervals is covered by a white square. Inside an interval, we have five tiling choices. We may cover the interval entirely with squares where the end points must be black and the interior, if it exists, must be white. Otherwise, we may cover the interval entirely with dominoes of the same color (when the interval has an even number of cells) or we may cover the interval with dominoes of the same color followed by a black square (when the interval has an odd number of cells). See Figure 9.8.
>
> When t is odd, we create intervals $I_1 = [a_2, a_3]$, $I_2 = [a_3, a_4], \dots, I_{(t-1)/2} = [a_{t-1}, a_t]$ that obey the same coloring rules as before. All cells outside these intervals are covered by a white square, except for cell a_1 which is covered by a black square. Since every interval allows five choices, the subset $\{a_1, \dots, a_t\}$ gives us $5^{\lfloor \frac{t}{2} \rfloor}$ ways to create a colored tiling.
>
> But not so fast. The coloring rules, as stated, have two deficiencies, which conveniently complement each other. The first problem is that a string of two or more dominoes of the same color can be generated by more than one subset. For example, the coloring in Figure 9.8 could also have been generated by the subset $\{3, 6, 8, \mathbf{9}, \mathbf{10}, 11, 12, 14, 15, 16, 18, 19\}$. The other problem is that the coloring rules

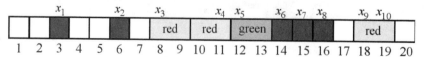

Figure 9.8. A colored tiling based on $S = \{3, 6, 8, 11, 12, 14, 15, 16, 18, 19\}$.

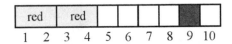

Figure 9.9. Another colored tiling.

provide no means of generating a colored tiling such as the one in Figure 9.9. We remedy these problems in one fell swoop by amending our coloring rules as follows. When an interval of even length $I_j = [a, b]$ is tiled by colored dominoes and I_j immediately precedes interval $I_{j+1} = [c, d]$, i.e., $c = b + 1$, then I_{j+1} can not be given the same color as I_j. Instead, we allow I_{j+1} to be covered by white squares ending with a single black square. Now the tiling in Figure 9.9 can *only* be obtained by the subset $\{1, 4, 5, 9\}$. Notice the amended rule still allows five choices for each interval, and that every subset $\{a_1, \ldots, a_t\}$ leads to exactly $5^{\lfloor \frac{t}{2} \rfloor}$ distinct n-tilings.

It is a little trickier to prove that every colored n-tiling is represented by exactly one subset S of $\{1, 2, \ldots, n\}$. Notice that if a tiling has no dominoes, then S is simply the set of cells covered by black squares. Otherwise, we can uniquely determine S by working backwards from the last domino and counting the number of black squares to the right of it. For more details, see [9]. For a different construction that generalizes to Lucas sequences of the first kind, see [48].

The same reasoning can be applied to the following Lucas identities. There are $2^n L_n$ ways to create a colored bracelet with the same types of tiles. As previously assumed, there are $L_0 = 2$ colored bracelets of size 0, one that is in-phase and the other out-of-phase.

Identity 236 $2^{n+1} f_n = \displaystyle\sum_{k=0}^{n} 2^k L_k.$

Set 1: The set of colored n-tilings. This set has size $2^n f_n$.

Set 2: The set of colored bracelets of size at most n. This set has size $\sum_{k=0}^{n} 2^k L_k$.

Correspondence: We establish a 1-to-2 correspondence between Set 1 and Set 2. Let T be a colored tiling of an n-board. If T does not consist of all white squares, let k denote the last cell covered by a nonwhite tile $(1 \leq k \leq n)$. After removing cells $k + 1$ through n, we generate two k-bracelets as illustrated in Figure 9.10:

B_1 An in-phase k-bracelet (ending with a nonwhite tile) obtained by gluing cells k and 1 together.

B_2 *If cell k is covered by a black square,* then B_2 is the in-phase k-bracelet obtained by replacing the kth cell of B_1 with a white square.

If cell k is covered by a colored domino, then B_2 is the out-of-phase k-bracelet obtained by rotating the tiles of B_1 clockwise one cell.

Every colored k-bracelet, where $1 \leq k \leq n$, is obtained exactly once in this fashion. The case where T consists of all white squares is identified with the two empty bracelets. Thus

$$2 \cdot 2^n f_n = \sum_{k=0}^{n} 2^k L_k.$$

Cell k covered by black square: **Cell k covered by colored domino:**

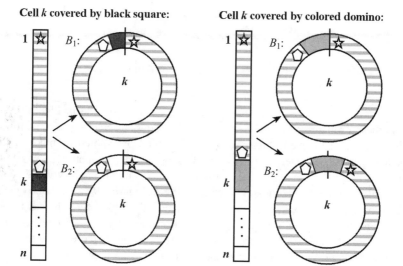

Figure 9.10. To prove that $2^{n+1}f_n = \sum_{k=0}^{n} 2^k L_k$, we draw a 1-to-2 correspondence depending on cell k, the last cell covered by a nonwhite tile.

Identity 237 $2\sum_{t=0}^{\lfloor \frac{n}{2} \rfloor} \binom{n}{2t} 5^t = 2^n L_n.$

Question: How many colored n-bracelets exist?

Answer 1: By definition, there are $2^n L_n$ such bracelets.

Answer 2: For each even subset $\{x_1, x_2, \ldots, x_{2t}\}$ of $\{1, 2, \ldots, n\}$, we shall generate $2 \cdot 5^t$ colored n-bracelets. As in the proof of Identity 235, create the intervals $I_1 = [x_1, x_2]$, $I_2 = [x_3, x_4], \ldots$, $I_k = [x_{2t-1}, x_{2t}]$. Then generate 5^t colored tilings of an n-board by following the coloring rules described there. Folding the n-board to become an n-bracelet, these colored tilings become *in-phase bracelets with an even number of black squares (possibly zero) before the first domino*. We call such a bracelet *simple*.

By the argument in Identity 235, all simple bracelets are generated by our coloring rules as $2t$ varies from 0 to n. To complete the identity we argue that simple bracelets account for exactly *half* of all possible colored bracelets. To see this, we draw a one-to-one correspondence between simple bracelets and the rest by conditioning on cell 1 of a simple bracelet. See Figure 9.11.

a If the first tile is a square, then change the color of the square covering cell 1 (producing an in-phase bracelet with an odd number of blacks before the first color).

b If the first tile is a domino, then rotate the simple bracelet counterclockwise one cell to produce a colored bracelet that is out-of-phase and therefore non-simple.

Thus there are as many simple colored bracelets as non-simple ones. Hence the total number of colored bracelets is

$$2\sum_{t=0}^{\lfloor \frac{n}{2} \rfloor} \binom{n}{2t} 5^t.$$

Simple bracelets: **Not simple bracelets:**

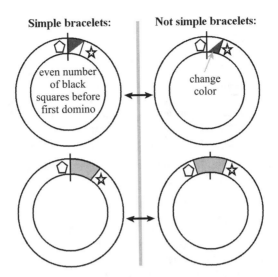

Figure 9.11. A 1-to-1 correspondence between simple bracelets (in-phase bracelets with an even number of black squares before the first domino) and not simple ones.

The combinatorial proofs given for the next two identities were discovered by David Gaebler and Robert Gaebler while they were undergraduates at Harvey Mudd College. Although they are special cases of Identity 11, we believe that the colored tiling correspondence is sufficiently novel to present here. We continue to color a square with one of two possible choices and color dominoes with one of four possible choices. But for ease of exposition, the domino colors will be either all black, all white, black-white, or white-black.

Identity 238 *For* $n \geq 1$, $f_{3n-1} = \sum_{k=1}^{n} \binom{n}{k} 2^k f_{k-1}.$

Set 1: The set of $(3n-1)$-tilings. This set has size f_{3n-1}.

Set 2: The set of pairs (X, B) where X is a nonempty subset of $\{1,\ldots,n\}$, and B is a colored j-tiling that begins with a square, where $j = |X|$. Conditioning on j, we have $\binom{n}{j}$ ways to choose X, f_{j-1} j-tilings that begin with a square, and 2^j ways to color each cell. There are $\sum_{j=1}^{n} \binom{n}{j} 2^j f_{j-1}$ such tilings.

Correspondence: Let A be a $(3n-1)$-tiling. We first insert a square tile to the right of cell $3k-1$ where $k \in \{1,\ldots,n\}$ is the smallest number for which A is breakable at cell $3k-1$. Call this new $(3n)$-tiling A'. See Figure 9.12.

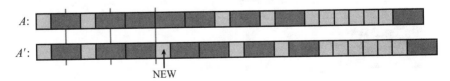

Figure 9.12. A $(3n-1)$-tiling becomes a $(3n)$-tiling by inserting a square at the first opportunity after a cell of the form $3k-1$.

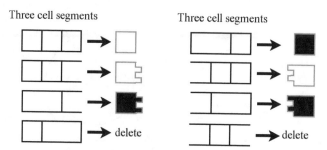

Figure 9.13. Every size 3 segment is transformed into a colored square or colored half-domino or is deleted.

By definition of k, A' begins with exactly $k-1$ "square-domino" pairs, and has a new square inserted at cell $3k$. (Cells $3k-1$ and $3k-2$ are covered by either a domino or two squares.) Next divide A' into n "segments" $s_1, s_2, \cdots, s_n$ of three cells each. This division will require true cutting as some segments could begin or end with a "half-domino". Now we convert it to a j-tiling for some $1 \le j \le n$ in 2-steps. For each of these segments, we transform it into a colored square or a colored half-domino or we delete it according to the transformation given in Figure 9.13.

Call the resulting configuration B', illustrated in Figure 9.14. B' is a collection of colored squares and half-dominoes occupying an n-board, but not all cells need to be covered. Finally we create B by "consolidating" B', removing all of the empty space. B will be a j-tiling where j is the number of segments that were not deleted. Let X denote the set of numbers i such that segment s_i was not deleted. Notice that our transformation always produces a legitimate colored tiling since every "left half-domino" is always completed with a "right half-domino." Also notice that by definition of k, segments $s_1, \ldots, s_{k-1}$ will be deleted since they are all of the form "square-domino", and segment s_k will be converted to a white square or black square. Thus B is guaranteed to begin with a square.

This procedure is easy to reverse. Given a colored j-tiling B that begins with a square, and a j-subset $X = \{x_1, x_2 \ldots, x_j\}$, we expand each square and half-

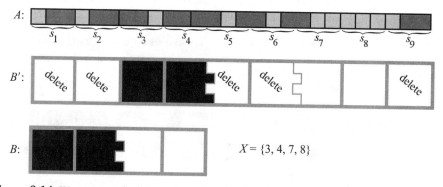

Figure 9.14. We convert a $(3n)$-tiling to a colored j-tiling by transforming each segment according to the rules of Figure 9.13. The size j subset X is the set of j undeleted segments.

domino into a length 3 segment by reversing the transformation of Figure 9.13. Then we create B' by placing these segments on a $(3n)$-board according to the instructions provided by X: For $1 \le i \le j$, the ith segment occupies cells $3i - 2, 3i - 1$, and $3i$. The unoccupied cells of B', in groups of three, are covered with either a "square-domino" or a "half-domino-square-half-domino", according to whether the empty interval is to be closed at both ends or open at both ends. Removing the colors gives us the $(3n)$-tiling A'. Since B began with a colored square, the tile on cell $3x_1$ of A' is guaranteed to be a square. Removing that square produces the original tiling A.

The next identity generalizes the previous one to Gibonacci numbers, discussed in Chapter 2. The proof is similar but with one little "twist".

Identity 239 $G_{3n} = \sum_{j=0}^{n} \binom{n}{j} 2^j G_j.$

Set 1: The set of phased $(3n)$-tilings, where an initial domino has G_0 phases and an initial square has G_1 phases. From Chapter 2, we know that this set has size G_{3n}.

Set 2: The set of pairs (X, B) where X is a (possibly empty) subset of $\{1, \ldots, n\}$, and B is a phased, colored j-tiling, where $j = |X|$. Here every cell, including the initial one, is assigned a color. Conditioning on j, we have $\binom{n}{j}$ ways to choose X, G_j phased j-tilings, and 2^j ways to color each cell. So this set has size $\sum_{j=0}^{n} \binom{n}{j} 2^j G_j$.

Correspondence: Let A be a phased $(3n)$-tiling. This time we locate the *last* breakable cell of the form $3k - 2$. If no such cell exists, then A must be of the form "domino-square-domino-square-$\cdots$-domino-square". There are G_0 such tilings, based on the phase of the initial domino. Otherwise, $k \ge 1$ and cells $3k - 1$ and $3k$ are covered either by two squares or by a single domino. Further, cells $3k + 1$ through $3n$ are covered by "domino-square-domino-square-$\cdots$-domino-square" ($n - k$ times). Next chop off the "tail" of the board after cell $3k - 2$, reverse these tiles and reattach it to the front end of the board. Call the new tiling A'. See Figure 9.15. Now break

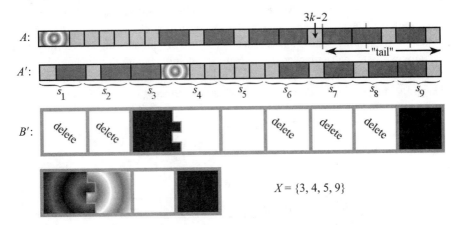

Figure 9.15. By a similar procedure, a phased $(3n)$-tiling becomes a phased colored j-tiling (for some $0 \le j \le n$) with an accompanying subset.

the board into n segments of three cells as in the previous proof, and using the transformation of Figure 9.13, create the configuration B' by turning each segment into a colored square, a colored half-domino, or deleting it. Notice that by construction, the first $n-k$ segments will all be "square-domino" and therefore deleted. The next segment will begin with the tile(s) formerly occupying cells $3k-1$ and $3k$ of A, followed by the phased square or half-domino that occupied the first cell of A. If the first tile of A was a square, then this segment will be transformed into a colored square and we assign it the same phase as A. If the first tile of A was a domino, then this segment will be transformed into a colored domino and we assign it the same phase as A. From B' we create j-subset X and phased j-tiling B exactly as in the previous proof. Like the procedure in the previous proof, it is also easily reversed.

Since Gibonacci numbers generalize Fibonacci numbers, Identity 239 generalizes Identity 238. In fact, we could prove Identity 238 directly by viewing f_{3n-1} as F_{3n}, the number of $(3n)$-tilings that begin with a square. Consequently, the resulting colored n-tiling will also begin with a colored square.

9.3   Some "Random" Identities and the Golden Ratio

Among the multitude of Fibonacci identities, we have not yet proved what may be the most important one of all. It is time to remedy that situation.

Identity 240 (Binet's Formula) *For $n \geq 0$,*

$$F_n = \frac{1}{\sqrt{5}}\left[\left(\frac{1+\sqrt{5}}{2}\right)^n - \left(\frac{1-\sqrt{5}}{2}\right)^n\right].$$

Here, F_n is the traditional definition of the Fibonacci number briefly discussed on page 10. But how in the world can one expect to find a combinatorial interpretation of such irrational quantities as $\sqrt{5}$ and $\phi = \frac{1+\sqrt{5}}{2}$? The answer: use probability.

To prove Binet's formula, we tile an infinite board by independently placing squares and dominoes, one after another. At each decision, we use a square with probability $1/\phi$ or a domino with probability $1/\phi^2$, where $\phi = \frac{1+\sqrt{5}}{2} \approx 1.618$. Conveniently, $1/\phi + 1/\phi^2 = 1$. A random example is shown in Figure 9.16. In this model, the probability that a tiling begins with any particular length n sequence of squares and dominoes is $1/\phi^n$. For example, the probability that a random tiling begins as in Figure 9.16 is $1/\phi^{12}$.

We apply this model to derive Binet's formula. Since $(1-\sqrt{5})/2 = -1/\phi$, and $f_n = F_{n+1}$, Identity 240 says

$$f_n = \frac{1}{\sqrt{5}}\left[\phi^{n+1} - \left(\frac{-1}{\phi}\right)^{n+1}\right]. \tag{9.1}$$

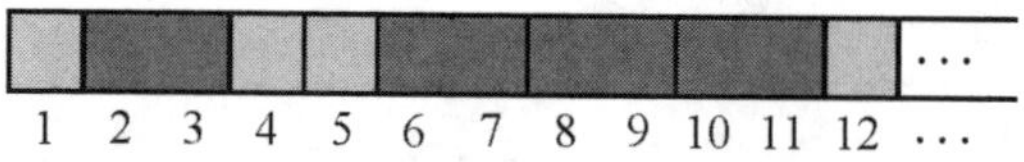

Figure 9.16. A random tiling of squares and dominoes.

Let q_n be the probability that a random tiling is breakable at cell n. Since there are f_n different ways to tile the first n cells,

$$q_n = \frac{f_n}{\phi^n}. \tag{9.2}$$

For a tiling to be unbreakable at n, it must be breakable at $n-1$ followed by a domino. Thus for $n \geq 1$, $1 - q_n = q_{n-1}/\phi^2$, or equivalently,

$$q_n = 1 - \frac{q_{n-1}}{\phi^2}, \tag{9.3}$$

where $q_0 = f_0 = 1$. Let $q = \lim_{n \to \infty} q_n$. If we assume, as will shortly be apparent, that the limit of (9.3) exists, then this limit q must satisfy $q = 1 - q/\phi^2$. Solving for q, we find that $q = (1 + 1/\phi^2)^{-1} = \phi/\sqrt{5}$.

Combined with (9.2), this gives us the asymptotic form of Binet's formula

$$f_n \approx \frac{\phi^{n+1}}{\sqrt{5}}.$$

To derive Binet's formula exactly, simply unravel recurrence (9.3) along with initial condition $q_0 = 1$ to get

$$q_n = 1 - \frac{1}{\phi^2} + \frac{1}{\phi^4} - \frac{1}{\phi^6} + \cdots + \left(\frac{-1}{\phi^2}\right)^n. \tag{9.4}$$

This is a finite geometric series (Identity 216 from Chapter 8) that simplifies to

$$q_n = \frac{\phi}{\sqrt{5}} \left[1 - \left(\frac{-1}{\phi^2}\right)^{n+1} \right]. \tag{9.5}$$

Thus by equation (9.2),

$$f_n = \phi^n q_n = \frac{\phi^{n+1}}{\sqrt{5}} \left[1 - \left(\frac{-1}{\phi^2}\right)^{n+1} \right] = \frac{1}{\sqrt{5}} \left[\phi^{n+1} - \left(\frac{-1}{\phi}\right)^{n+1} \right],$$

as desired.

In fact, Binet's formula can be simplified as follows.

Corollary 30 *For $n \geq 0$, f_n is the integer closest to $\phi^{n+1}/\sqrt{5}$.*

Proof. This is equivalent to saying that $|f_n - \phi^{n+1}/\sqrt{5}| < \frac{1}{2}$. By Binet's formula this amounts to showing that $\sqrt{5}\phi^{n+1} > 2$, which is clearly true for all $n \geq 0$. $\diamond$

Binet's formula implies that consecutive Fibonacci numbers essentially grow by a factor of ϕ. More precisely,

Corollary 31 *For $n, m \geq 0$,*

$$\lim_{n \to \infty} \frac{f_{n+m}}{f_n} = \phi^m.$$

Proof. By examining the limiting value of q_n given in the derivation of Binet's formula, we have for any $m \geq 0$,

$$\phi/\sqrt{5} = \lim_{n \to \infty} \frac{f_n}{\phi^n} = \lim_{n \to \infty} \frac{f_{n+m}}{\phi^{n+m}}.$$

Thus,

$$1 = \lim_{n \to \infty} \frac{f_{n+m}/\phi^{n+m}}{f_n/\phi^n} = \lim_{n \to \infty} \frac{f_{n+m}}{f_n \phi^m},$$

hence, $\lim_{n \to \infty} f_{n+m}/f_n = \phi^m$.   ◇

Other identities involving Fibonacci numbers and the golden ratio are also obtainable by our probabilistic approach. For instance, by substituting equation (9.2) into equation (9.3) and multiplying by ϕ^n we have also demonstrated

Corollary 32 *For $n \geq 1$,*

$$\phi^n = f_n + f_{n-1}/\phi.$$

And since $f_n = f_{n-1} + f_{n-2} = f_{n-1}(\phi - \frac{1}{\phi}) + f_{n-2}$, we also have

Corollary 33 *For $n \geq 1$,*

$$\phi^n = \phi f_{n-1} + f_{n-2}.$$

By equation (9.4), we have

$$q_n - q_{n-1} = (-1/\phi^2)^n.$$

For a direct probabilistic proof of this, see [5]. Multiplying this by ϕ^n gives us

Corollary 34 *For $n \geq 1$,*

$$f_n - \phi f_{n-1} = \frac{(-1)^n}{\phi^n}.$$

Dividing this last equality by f_{n-1} gives us for all $n \geq 1$,

$$f_n/f_{n-1} - \phi = \frac{(-1)^n}{\phi^n f_{n-1}},$$

which demonstrates that successive ratios of Fibonacci numbers get closer and closer to ϕ.

For other proofs of Binet's formula and other Fibonacci identities using probability see [5] and [4].

The same probabilistic argument can be used to derive a combinatorial proof of Binet's formula for Lucas numbers.

Identity 241 *For $n \geq 0$,*

$$L_n = \left(\frac{1 + \sqrt{5}}{2} \right)^n + \left(\frac{1 - \sqrt{5}}{2} \right)^n.$$

Here, we tile an infinite board one tile at a time, beginning at cell 1. The first tile is either a square with probability $1/\sqrt{5}$, an in-phase domino with probability $1/(\phi\sqrt{5})$ or an out-of-phase domino with probability $1/(\phi\sqrt{5})$. Thereafter, tiles are chosen at random and independently with probability $1/\phi$ for squares and $1/\phi^2$ for dominoes. In this model, any length n tiling has probability $1/(\phi^{n-1}\sqrt{5})$. Let r_n denote the probability that a tiling is breakable at $n+1$. Thus for $n \geq 2$,

$$r_n = \frac{L_n}{\phi^{n-1}\sqrt{5}},$$

where $r_1 = 1/\sqrt{5}$. By the same argument as Identity 240, for $n \geq 2$,

$$r_n = 1 - \frac{r_{n-1}}{\phi^2},$$

which unravels to

$$r_n = 1 - \frac{1}{\phi^2} + \frac{1}{\phi^4} - \frac{1}{\phi^6} + \cdots + \left(\frac{-1}{\phi^2}\right)^{n-2} + \frac{(-1/\phi^2)^{n-1}}{\sqrt{5}}.$$

Summing the series results in

$$r_n = \frac{\phi}{\sqrt{5}}\left[1 + \left(\frac{-1}{\phi^2}\right)^n\right]$$

which is the same as our original identity with both sides divided by $\phi^{n-1}\sqrt{5}$.

As we did for Fibonacci numbers, we have an immediate corollary.

Corollary 35 *For $n \geq 2$, L_n is the integer closest to ϕ^n.*

Notice that if we rewrite Binet's formula as

$$\sqrt{5}F_n = \phi^n - (-1/\phi)^n,$$

and add and subtract it to the previous identity

$$L_n = \phi^n + (-1/\phi)^n,$$

we get the so-called deMoivre theorems.

Identity 242 *For $n \geq 0$,*
$$\phi^n = \frac{\sqrt{5}F_n + L_n}{2}.$$

Identity 243 *For $n \geq 0$,*
$$\left(\frac{-1}{\phi}\right)^n = \frac{L_n - \sqrt{5}F_n}{2}.$$

With just a little bit more work, we can derive a Binet-like formula for Gibonacci numbers.

Identity 244 *For $n \geq 0$, $G_n = \alpha\phi^n + \beta(-1/\phi)^n$, where $\alpha = (G_1 + G_0/\phi)/\sqrt{5}$ and $\beta = (\phi G_0 - G_1)/\sqrt{5}$.*

We prove this using a random tiling model similar to the previous ones for Fibonacci and Lucas numbers. Consider tiling an infinite board, one tile at a time, beginning at cell 1. The first tile is either a phased square or phased domino and is chosen in a random manner to be described in the next paragraph. Thereafter, tiles are chosen at random and independently with probability $1/\phi$ for squares and $1/\phi^2$ for dominoes. Our goal is to assign the initial probabilities in such a way that any particular tiling of the first n cells has probability p_n of occurring, where p_n depends only on n.

To achieve this goal, let P_d and P_s denote the probability that a tiling begins with a domino or square respectively. For a tiling beginning with a domino, its phase is chosen randomly from among the G_0 possible phases. Likewise, a tiling beginning with a square has its phase chosen randomly from among the G_1 possible phases. Hence the probability that a random tiling begins with a particular phased domino or square is P_d/G_0 and P_s/G_1 respectively. Since we desire that a particular phased square followed by an unphased square should have the same probability p_2 as a particular phased domino, we must have

$$\frac{P_s}{G_1}\frac{1}{\phi} = \frac{P_d}{G_0}.$$

Combined with $P_d + P_s = 1$, it follows that

$$P_d = \frac{G_0}{G_0 + \phi G_1}, \quad P_s = \frac{\phi G_1}{G_0 + \phi G_1}.$$

Consequently, the probability of beginning with any particular n-tiling is

$$p_n = \frac{1}{G_1}P_s\frac{1}{\phi^{n-1}} = \frac{1}{(G_0 + \phi G_1)\phi^{n-2}}.$$

If we let r_n denote the probability that a random tiling is breakable at cell n, we have $r_n = G_n p_n$. In other words, for $n \geq 1$,

$$G_n = r_n(G_0 + \phi G_1)\phi^{n-2}. \tag{9.6}$$

Next we compute r_n directly. Notice that r_n must satisfy for $n \geq 2$,

$$1 - r_n = r_{n-1}\frac{1}{\phi^2},$$

since a tiling is unbreakable at cell n if and only if it was breakable at cell $n-1$ followed by a domino. Unraveling this recurrence we get

$$r_n = 1 - r_{n-1}\frac{1}{\phi^2}$$

$$= 1 - \frac{1}{\phi^2} + \frac{1}{\phi^4}r_{n-2}$$

$$= 1 - \frac{1}{\phi^2} + \frac{1}{\phi^4} - \frac{1}{\phi^6}r_{n-3}$$

$$\vdots$$

$$= 1 - \frac{1}{\phi^2} + \frac{1}{\phi^4} - \frac{1}{\phi^6} + \cdots + \left(\frac{-1}{\phi^2}\right)^{n-2} + \left(\frac{-1}{\phi^2}\right)^{n-1}r_1.$$

As

$$r_1 = P_s = \frac{\phi G_1}{G_0 + \phi G_1},$$

and the previous terms can be summed as a finite geometric series, we get (after a bit of algebra including $\phi^2 + 1 = \phi\sqrt{5}$)

$$r_n = \frac{\phi}{\sqrt{5}} + \frac{(-1)^n}{\phi^{2n-2}} \cdot \left(\frac{\phi G_0 - G_1}{\sqrt{5}(G_0 + \phi G_1)} \right) = \frac{\phi}{\sqrt{5}} + \frac{(-1)^n \beta}{\phi^{2n-1}\sqrt{5}\alpha}.$$

Substituting this into equation (9.6), gives us the desired identity.

In fact, Binet-like formulas for most kth order recurrences can be probabilistically derived using Markov chains, as is done in [4]. For probabilistic proofs of other Fibonacci identities such as

$$\sum_{n \geq 1} \frac{F_n}{2^n} = 2,$$

$$\sum_{n \geq 1} n\frac{F_n}{2^n} = 10,$$

$$\sum_{n \geq 1} n^2\frac{F_n}{2^n} = 94,$$

see [5] and [7].

9.4   Fibonacci and Lucas Polynomials

Another natural generalization of Fibonacci numbers are the *Fibonacci polynomials* defined recursively by $f_0(x) = F_1(x) = 1$, $f_1(x) = F_2(x) = x$ and for $n \geq 2$,

$$f_n(x) = xf_{n-1}(x) + f_{n-2}(x).$$

For example, $f_2(x) = x^2 + 1$, $f_3(x) = x^3 + 2x$, $f_4(x) = x^4 + 3x^2 + 1$, $f_5(x) = x^5 + 4x^3 + 3x$, and so on. Notice that when $x = 1$, the initial conditions and recurrence simply generate the Fibonacci numbers, i.e., $f_n(1) = f_n$. Inductively, it is clear that $f_n(x)$ is an nth degree polynomial and therefore has the form

$$f_n(x) = \sum_{k=0}^{n} f(n, k)x^k.$$

Naturally, $f(n, k)$ must be counting something. By conditioning on the last tile, we leave it to the reader to verify

Combinatorial Theorem 12 *The Fibonacci polynomial $f_n(x) = \sum_{k=0}^{n} f(n, k)x^k$, where $f(n, k)$ counts n-tilings with exactly k squares.*

Since an n-tiling with exactly k squares must have $(n - k)/2$ dominoes and therefore $(n + k)/2$ tiles, it follows that

$$f(n, k) = \binom{(n + k)/2}{k},$$

which equals 0 when n and k have opposite parity.

Many Fibonacci number identities have analogous Fibonacci polynomial identities. We list just two examples and invite the reader to explore more of their own.

Identity 245 *For* $m, n \geq 1$,

$$f_{m+n}(x) = f_m(x)f_n(x) + f_{m-1}(x)f_{n-1}(x).$$

Question: For each $k \geq 0$, how many $(m + n)$-tilings have exactly k squares?

Answer 1: $f(m + n, k)$, that is, the x^k coefficient of $f_{m+n}(x)$.

Answer 2: First we count those tilings that are breakable at cell m. Such a tiling must have $0 \leq j \leq m$ squares in the m-tiling covering cells 1 through m and $k - j$ squares in the n-tiling covering cells $m + 1$ through $m + n$. Hence the number of breakable tilings with k squares is

$$\sum_{j=0}^{m} f(m, j)f(n, k - j).$$

By similar reasoning, the number of unbreakable tilings with k squares is

$$\sum_{j=0}^{m-1} f(m - 1, j)f(n - 1, k - j).$$

Consequently for all $k \geq 0$

$$f(m + n, k) = \sum_{j=0}^{m} f(m, j)f(n, k - j) + \sum_{j=0}^{m-1} f(m - 1, j)f(n - 1, k - j),$$

which is precisely the way one would compute the x^k coefficient of $f_m(x)f_n(x) + f_{m-1}(x)f_{n-1}(x)$.

For the next identity, we observe that the familiar tail swapping technique, first exploited in Identity 8 from Chapter 1, preserves the total number of squares in the tiling pair. That is, tail swapping provides a bijection between tiling pairs with the same number of squares (with the lone exception of when there are 0 squares in the tiling pair.)

Identity 246 *For* $n \geq 1$,

$$f_n^2(x) - f_{n-1}(x)f_{n+1}(x) = (-1)^n.$$

In a similar fashion, one can define *Lucas polynomials* by $L_0(x) = 2$, $L_1(x) = x$, and for $n \geq 2$, $L_n(x) = xL_{n-1}(x) + L_{n-2}(x)$. More generally, *Gibonacci polynomials* can be defined by $G_0(x) = G_0$, $G_1(x) = G_1 x$, and for $n \geq 2$, $G_n(x) = xG_{n-1}(x) + G_{n-2}(x)$. Just as with Fibonacci polynomials, we have

Combinatorial Theorem 13 *The Gibonacci polynomial* $G_n(x) = \sum_{k=0}^{n} G(n, k)x^k$, *where* $G(n, k)$ *counts phased* n-*tilings with exactly* k *squares, where an initial domino has* G_0 *phases and an initial square has* G_1 *phases.*

Naturally, these ideas can be extended to higher order recurrences and colored tilings, and we invite the reader to do so.

9.5 Negative Numbers

Practically all of the identities in this book have had two things in common. The objects that we have been counting have been nonnegative integer quantities, typically with a nonnegative integer index. In Chapter 3 we introduced weighted tilings as a way to understand linear recurrences with arbitrary coefficients and initial conditions. However, we have not examined the "negative 7th" Gibonacci number G_{-7}, even though we could compute it, given arbitrary initial conditions G_0, G_1, and the Fibonacci recurrence $G_n = G_{n-1} + G_{n-2}$ for *all* integers n. And yet, many of the identities proved in this book remain true even for negative indices. It is easy to prove by induction that

$$F_{-n} = (-1)^{n+1} F_n \quad \text{or} \quad f_{-n} = (-1)^n f_{n-2},$$

$$L_{-n} = (-1)^n L_n,$$

and

$$G_{-n} = (-1)^n H_n,$$

where H_n is the Gibonacci sequence defined by $H_0 = G_0$, $H_1 = G_0 - G_1$, and for $n \geq 2$, $H_n = H_{n-1} + H_{n-2}$. These could almost serve as definitions for negatively indexed Fibonacci, Lucas, and Gibonacci numbers. Propp [45] approaches this subject by counting "signed matchings" on a $2 \times n$ grid. There is a remarkable theorem by Bruckman and Rabinowitz [18] that shows if an identity involving numbers generated by a second order recurrence holds for all positive subscripts, then it also holds for negative subscripts. Is there a natural *combinatorial* interpretation of generalized Fibonacci numbers that allow us to understand identities such as

$$G_{m+n} = G_m f_n + G_{m-1} f_{n-1}$$

just as easily when m or n are negative? We leave that question for future consideration.

9.6 Open Problems and Vajda Data

Instead of ending this chapter with exercises, we leave the reader with some open problems. We hope that the reader has been convinced of the power and simplicity of combinatorial proofs, particularly for identities concerning Fibonacci numbers and their generalizations.

To indicate the power of our approach, we refer to the classic book *Fibonacci & Lucas Numbers and the Golden Section* by Steven Vajda [58], which contains 118 identities involving Fibonacci, Lucas, and Gibonacci numbers. These identities are proved by a myriad of algebraic methods—induction, generating functions, hyperbolic functions, to name a few. Although *none* are proved combinatorially in the book, we have used tiling to explain 91 of these identities—and counting!

We leave the reader with a list of those 27 identities from [58] that have thus far resisted combinatorial explanation as far as we know. Some of the original identities have been restated (e.g., $F_n = f_{n-1}$ and other re-indexing) for combinatorial clarity.

This first identity is easily derived by elementary algebra using $\phi - 1/\phi = 1$ and $G_0 + G_1 = G_2$, but is there a combinatorial (probabilistic?) explanation?

V57:

$$(G_0/\phi + G_1)(G_0\phi - G_1) = G_0 G_2 - G_1^2.$$

Vajda identities V69 through V76 seem awfully similar. One good idea might solve them all!

V69:

$$\sum_{i=0}^{2n} \binom{2n}{i} f_{2i-1} = 5^n f_{2n-1}.$$

V70:

$$\sum_{i=0}^{2n+1} \binom{2n+1}{i} f_{2i-1} = 5^n L_{2n+1}.$$

V71:

$$\sum_{i=0}^{2n} \binom{2n}{i} L_{2i} = 5^n L_{2n}.$$

V72:

$$\sum_{i=0}^{2n+1} \binom{2n+1}{i} L_{2i} = 5^{n+1} f_{2n}.$$

V73:

$$\sum_{i=0}^{2n} \binom{2n}{i} f_{i-1}^2 = 5^{n-1} L_{2n}.$$

V74:

$$\sum_{i=0}^{2n+1} \binom{2n+1}{i} f_{i-1}^2 = 5^n f_{2n}.$$

V75:

$$\sum_{i=0}^{2n} \binom{2n}{i} L_i^2 = 5^n L_{2n}.$$

V76:

$$\sum_{i=0}^{2n+1} \binom{2n+1}{i} L_i^2 = 5^{n+1} f_{2n}.$$

Perhaps the next one can be explained using probability?

V77:

$$\sum_{i\geq 1} \frac{1}{F_{2^i}} = 4 - \phi = 3 - \frac{1}{\phi}.$$

Perhaps the techniques of Chapter 6 can be applied to Vajda identities V78–V88?

V78:

$$L_t^{2k+1} = \sum_{i=0}^{k} \binom{2k+1}{i} (-1)^{it} L_{(2k+1-2i)t}.$$

V79:

$$L_t^{2k} = \binom{2k}{k}(-1)^{tk} + \sum_{i=0}^{k-1}\binom{2k}{i}(-1)^{it}L_{(2k-2i)t}.$$

V80:

$$5^k F_t^{2k+1} = \sum_{i=0}^{k}\binom{2k+1}{i}(-1)^{i(t+1)}F_{(2k+1-2i)t}.$$

V81:

$$5^k f_{t-1}^{2k} = \binom{2k}{k}(-1)^{(t+1)k} + \sum_{i=0}^{k-1}\binom{2k}{i}(-1)^{i(t+1)}L_{(2k-2i)t}.$$

V82:

$$L_{kt} = L_t^k + \sum_{i=1}^{k/2}\frac{k}{i}(-1)^{i(t+1)}L_t^{k-2i}\binom{k-i-1}{i-1}.$$

V83:

$$F_{(2k+1)t} = 5^k F_t^{2k+1} + \sum_{i=1}^{k}\frac{2k+1}{i}(-1)^{it}5^{k-i}\binom{2k-i}{i-1}F_t^{2k+1-2i}.$$

V84:

$$L_{2kt} = 5^k F_t^{2k} + \sum_{i=1}^{k}\frac{2k}{i}(-1)^{it}5^{k-i}\binom{2k-i-1}{i-1}F_t^{2k-2i}.$$

V85: For $k \geq 0$,

$$F_{(2k+3)t} = (-1)^{(k+1)t} + F_t\sum_{i=0}^{k}(-1)^{it}L_{(2k+2-2i)t}.$$

V86: For $k \geq 0$,

$$F_{(2k+2)t} = F_t\sum_{i=0}^{k}(-1)^{it}L_{(2k+1-2i)t}.$$

V87: For $k \geq 0$,

$$L_{(2k+3)t} = (-1)^{(k+1)(t+1)} + L_t\sum_{i=0}^{k}(-1)^{i(t+1)}L_{(2k+2-2i)t}.$$

V88: For $k \geq 0$,

$$F_{(2k+2)t} = L_t\sum_{i=0}^{k}(-1)^{i(t+1)}F_{(2k+1-2i)t}.$$

V89:

$$\sum_{i=1}^{n}\frac{(-1)^{2^{i-1}r}}{F_{2ir}} = \frac{(-1)^r F_{(2^n-1)r}}{F_r F_{2^n r}}.$$

This next identity seems to be asking for a probabilistic argument.

V90:

$$\frac{1}{F_4} + \frac{1}{F_8} + \cdots + \frac{1}{F_{2^n}} = 1 - \frac{F_{2^n - 1}}{F_{2^n}}.$$

V93:

$$5 \sum_{i=1}^{n+1} (-1)^{ir} F_{ir}^2 = (-1)^{(n+1)r} \frac{F_{(2n+3)r}}{F_r} - 2n - 3.$$

V94:

$$\sum_{i=1}^{n+1} (-1)^{ir} L_{ir}^2 = (-1)^{(n+1)r} \frac{F_{(2n+3)r}}{F_r} + 2n + 1.$$

Perhaps this next one can be done by a probabilistic argument?

V103:

$$\sum_{n \geq 0} \frac{(-1)^n}{f_{n+1} f_n} = \frac{1}{\phi}.$$

And finally, a continued fraction identity.

V106:

$$\frac{F_{(t+1)m}}{F_{tm}} = L_m - \cfrac{(-1)^m}{L_m - \cfrac{(-1)^m}{L_m - \cfrac{(-1)^m}{\ddots - \cfrac{(-1)^m}{L_m}}}},$$

where the number L_m appears t times.

We have every confidence that all of the above identities will be combinatorially explained someday. You can count on it!

Some Hints and Solutions for Chapter Exercises

Chapter 1

Identity 12. Condition on the location of the last square.

Identity 13. Condition on the breakability of a $(2n + 2)$-tiling at cell $n + 1$.

Identity 14. Count the number of pairs of n-tilings where at least one ends in a square. Condition on whether the first tiling ends in a square or not.

Identity 15. Same strategy as Identity 14.

Identity 16. Every n-tiling generates two tilings of size $n + 1$ or $n - 2$. For the first copy, attach a square to create all $(n + 1)$-tilings ending with a square. The second copy will depend on whether the last tile is a square or a domino.

Identity 17. Every n-tiling generates three tilings of size $n + 2$ or $n - 2$. For the first two copies, attach a domino or two squares. The third copy will depend on whether the last tile is a square or a domino.

Identity 18. Follows immediately from Identity 17 (the fourth copy remains unchanged!)

Identity 19. The second equality follows immediately from Identity 14. It remains to find a correspondence between the left side and the right side. The left side counts tiling 4-tuples (A, B, C, D) or (E, F, G, H) where A, B, C, D, E, F, G, H have respective sizes $n, n, n + 1, n + 1, n - 1, n - 1, n + 2, n + 2$. The right side counts (X, Y) where X and Y are both $(2n + 2)$-tilings. Each (A, B, C, D) generates four tiling pairs, the first three of which are: (AdB, CD), (CD, AdB), (CsA, DsB). These three tiling pairs cover all tiling pairs (X, Y) where X or Y (but not both) are unbreakable at cell $n + 1$ or X and Y are both breakable at cell $n + 1$, and both X and Y have cell $n + 2$ occupied by a square. Our fourth tiling pair generated by (A, B, C, D) has four cases, and conditions on how C and D end. Likewise, (E, F, G, H) generates one tiling pair and has four cases, depending on how G and H end. To reverse the procedure, consider the tiles that occupy cells $n + 1$ and $n + 2$.

We note that any triple of numbers (a, b, c) where $a = x^2 - y^2$, $b = 2xy$, $c = x^2 + y^2$ form a Pythagorean triple. This identity is just the special case where $a = f_{n+1}$ and $b = f_n$.

Identity 20. Condition on the number of dominoes that appear among the first p tiles. Given an initial segment of i dominoes and $p - i$ squares, $\binom{p}{i}$ counts the number of ways to select the i positions for the dominoes among the first p tiles. This initial segment has length $2i + (p - i) = p + i$. The rest of the board has length $(n + p) - (p + i) = n - i$ and can be covered f_{n-i} ways.

Identity 21. When n is even, Identity 21 unravels to become

$$f_0 + f_2 + \cdots + f_n = 1 + f_{n-1} + f_1 + f_3 + \cdots + f_{n-1},$$

which we prove here. We leave the odd case to you.

Question: For n even, in how many ways can an $(n+1)$-board be tiled using squares and dominoes?

Answer 1: Condition on the last square. Since $n + 1$ is odd, a last square must exist on an odd cell. For $0 \le 2k + 1 \le n + 1$, the number of $(n+1)$-tilings with last square on cell $2k + 1$ is f_{2k}. Altogether, the number of $(n+1)$-tilings is $f_0 + f_2 + \cdots + f_n$.

Answer 2: There are f_{n-1} such tilings that begin with a domino. Among those that begin with a square, we condition on the last square. There is 1 tiling consisting of a single square followed by all dominoes. For $3 \le 2k + 1 \le n + 1$, the number of $(n+1)$-tilings that begin with a square and whose last square occurs at cell $2k + 1$ is f_{2k-1}. Altogether, we have $f_{n-1} + 1 + f_1 + f_3 + \cdots + f_{n-1}$ tilings. See Figure HS.1.

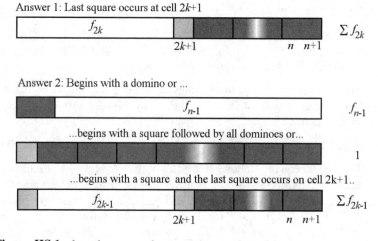

Figure HS.1. $f_0 + f_2 + \cdots + f_n = 1 + f_{n-1} + f_1 + f_3 + \cdots + f_{n-1}$ for n even.

Identity 22. Use Identity 21 with telescoping sums.

Identity 23. Question: How many $(3n + 2)$-tilings have their last domino ending on a cell of the form $3j + 2$ for $0 \le j \le n$?

Answer 1: A tiling with its last domino ending on cell $3j + 2$ begins with one of f_{3j} tilings, followed by a domino, and completed with all squares. Since j ranges from 0 to n, there are a total of $f_0 + f_3 + f_6 + \cdots + f_{3n}$ such tilings.

Answer 2: We demonstrate a one-to-one correspondence between the set of $(3n+2)$-tilings with last domino ending on cells of the form $3j + 2$ and the set of $(3n + 2)$-tilings with last domino ending on cells of the form $3j$ or $3j + 1$. If for some j,

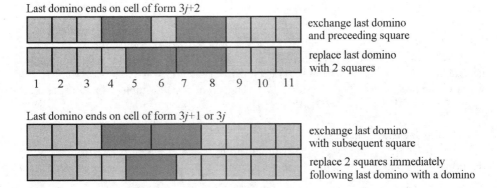

Last domino ends on cell of form $3j+2$

exchange last domino and preceeding square

replace last domino with 2 squares

Last domino ends on cell of form $3j+1$ or $3j$

exchange last domino with subsequent square

replace 2 squares immediately following last domino with a domino

Figure HS.2. A one-to-one correspondence between $(3n + 2)$-tilings with last domino ending on cells of the form $3j + 2$ and $(3n + 2)$-tilings with last domino ending on cells of the form $3j$ or $3j + 1$. So half of all $(3n + 2)$-tilings must have a domino that ends on a cell of form $3j + 2$ for $0 \le j \le n$.

$0 \le j \le n$, the last domino ends on cell $3j+2$ and is preceded by a square, exchange the square and the domino to give a tiling whose last domino ends on cell $3j + 1$. If the last domino ends on cell $3j + 2$ and is not preceded by a square (i.e., it is preceded by a domino or covers cells 1 and 2), replace the domino with two squares to give a tiling whose last domino ends on cell $3j$. See Figure HS.2. This process is completely reversible, so we see that half of all f_{3n+2} $(3n + 2)$-tilings have their last domino ending on cell $3j + 2$ for $0 \le j \le n$.

Identity 24. Argument is similar to Identity 23 except that we exclude the all square tiling from the right-hand side.

Identity 25. Argument is similar to Identity 23 except that we exclude the all square tiling from the right-hand side.

Identity 26. Tail trashing technique. Let X be a $(2n)$-tiling and Y a $(2n+1)$-tiling. If the last common fault occurs at an even cell, say $2j$, then $X = Ad^{n-j}$ and $Y = Bsd^{n-j}$. Generate the $(4j)$-tiling AB which is breakable at cell $2j$. Otherwise the last common fault occurs at an odd cell, say $2j - 1$. Here $X = Asd^{n-j}$ and $Y = Bd^{n-j+1}$. Generate the $(4j)$-tiling AdB which is unbreakable at cell $2j$.

Identity 27. Same tail trashing technique as in Identity 26.

Identity 28. Same tail trashing technique as in Identity 26.

Identity 29. Let X be a $(2n - 1)$-tiling covering cells 2 through $2n$ and let Y a $(2n + 1)$-tiling covering cells 1 through $2n + 1$. Proceed with tail trashing as in Identity 26.

Identity 31. The left side counts 4-tuples of n-tilings. The product on the right side counts 4-tuples (W, X, Y, Z), where W, X, Y, Z are tilings with respective lengths $n + 2, n - 2, n + 1, n - 1$, and begin on respective cells $1, 3, 1, 2$. (X is centered under W and Z is centered under Y.)

We first consider the case where n is even. Here, the (Y, Z) pair must contain a fault. If (W, X) also has a fault, then tail swapping both pairs produces (W', X', Y', Z')

which is a 4-tuple of n-tilings. When (W, X) is fault-free, (when $W = sd^{n/2}s$, $X = d^{(n-2)/2}$), we produce $(Y', Z', d^{n/2}, d^{n/2})$. This produces all 4-tuples of n-tilings except $(d^{n/2}, d^{n/2}, d^{n/2}, d^{n/2})$.

When n is odd, the (W, X) pair must have a fault. If (Y, Z) also has a fault, we again generate (W', X', Y', Z'). When (Y, Z) is fault-free (when $Y = d^{(n+1)/2}$, $Z = d^{(n-1)/2}$), we associate $(sd^{(n-1)/2}, d^{(n-1)/2}s, W', X')$. This produces all 4-tuples of n-tilings except $(sd^{(n-1)/2}, d^{(n-1)/2}s, sd^{(n-1)/2}, d^{(n-1)/2}s)$.

Combinatorial Interpretations

1. For each $(n + 1)$-tiling create the n-tuple $(b_1, b_2, \ldots, b_n)$ where $b_i = 0$ if and only if the tiling is unbreakable at cell i.

2 For $1 \leq i \leq n$, i is in S if and only if the tiling is unbreakable at cell i.

3. Let T be a tiling of an n-board as described in the problem. For each tile of length $\ell \geq 2$, break the tile into a domino followed by $\ell - 2$ squares. Then remove the first domino.

4. Let T be a tiling of an n-board as described in the problem. Break each odd length tile into a square followed by dominoes. Remove the first square.

5. For each arrangement $a_1 a_2 \ldots a_n$, create the n-tiling where cell i is covered by a square if and only if $a_i = i$.

6. Our ternary sequence is of the form $2^{a_0} 0^{b_0} 1 2^{a_1} 0^{b_1} 1 \cdots 2^{a_j} 0^{b^j} 1^c$, where $a_i, b_i \geq 0$, and $c_i \in \{0, 1\}$. Reading from left to right, each 2 becomes a domino, each 1 becomes a square, and for $1 \leq i \leq j$, 0^{b_i} becomes sd^{b_i}. If $c = 1$, it becomes two squares instead of one.

7. Let X be an n-tiling described in the hint. If a tile of length m covers cells $k + 1$ through $k + m$ with highlighted cell $k + i$, then in the square-domino $(2n - 1)$-tiling Y, we cover cells $2k + 1$ through $2k + m - 1$ with $m - 1$ dominoes and 1 square, with the square on cell $2i - 1$. Unless this tile is the last tile of X, cell $2k + 2m$ of Y is covered with a square. Observe that the jth tile of X ends at cell k if and only if the $2j$th square of Y is at cell $2k$.

8. Here X counts n-tilings where tiles can be of any length, and squares are assigned a color: black or white. Y consists of all $(2n)$-tilings with squares and dominoes. Each tile of X is "doubled" as follows. Black squares in X become dominoes in Y. For all other tiles, a tile of length k becomes $sd^{k-1}s$, a square followed by $k - 1$ dominoes followed by a square.

9. Let $b_0 = 1$ and given the sequence $(b_1, \ldots, b_n)$ create the $(n + 1)$-tiling where cell i has a square if and only if $b_i \neq b_{i-1}$.

Chapter 2

Identity 50. Condition on the tile covering cell 1.

Identity 51. Each n-tiling generates two objects. The second object depends on whether the tiling begins with a square or a domino.

Identity 52. Every n-tiling generates five bracelets of length n or $n + 1$. These bracelets depend on whether they started with a square or a domino. When the dust settles every

$(n+1)$-bracelet will be generated only once and every n-bracelet will be generated twice. For the first three bracelets, create two copies of an in-phase n-bracelet and one copy of an in-phase $(n + 1)$-bracelet that begins with a square. The other two bracelets generated by the n-tiling will depend on how the n-tiling begins.

Identity 53. Offset the n-tiling pair by 1, then tail swap (if possible) to create the ordered pair (A, B), where A is an $(n+1)$-tiling and B is an $(n-1)$ tiling. From these, we generate five $(n+1)$-bracelet pairs. The first four of them are $(A, ssB), (A, dB), (A, d^-B), (d^-B, A)$, where all of these bracelets are in-phase except for those that begin with d^-. The fifth bracelet will depend on how A and B begin.

Identity 54. The proof is similar to the proof of Identity 234.

Identity 55. Rewrite the identity as $L_{2n+1} + L_{2n-3} + L_{2n-7} + \cdots = f_{2n+1} + L_{2n-1} + L_{2n-5} + \cdots$. How many tilings (not bracelets) of odd length less than or equal to $2n + 1$ exist? Exploit the fact that $f_m + f_{m-2} = L_m$. For example $(f_9 + f_7) + (f_5 + f_3) + f_1 = L_9 + L_5 + f_1 = L_9 + L_5 + L_1$. Also $f_9 + (f_7 + f_5) + (f_3 + f_1) = f_9 + L_7 + L_3$.

Identity 56. This is a special case of Identity 59.

Identity 57. L_n counts the number of phased n-tilings where an initial domino is either in-phase or out-of-phase. (We shall still refer to these as n-bracelets.) For each n-bracelet X, generate $n + 1$ bracelet pairs (whose lengths sum to n) as follows. For each cell $1 \leq j \leq n - 1$, if X is breakable at cell j, then generate the natural j-bracelet (with the same phase as X) and an in-phase $(n - j)$-bracelet. Otherwise, generate the natural $(j - 1)$-bracelet (with the same phase as X) and an out-of-phase $(n - j + 1)$-bracelet. Cell n generates two bracelet pairs $(X, \emptyset^+)$ and $(X, \emptyset^-)$. This process generates all but $2f_n$ bracelet pairs. The missing bracelet pairs are $(\emptyset^+, Y)$ and $(\emptyset^-, Y)$, where Y is an in-phase n-bracelet. (Due to Dan Cicio.)

Identity 58. For every n-bracelet X, we usually generate n tiling pairs whose lengths sum to $n - 2$. The f_{n-1} exceptions occur for in-phase bracelets that end with a square, where we only generate $n - 1$ tiling pairs. Every tiling pair will be generated five times. Specifically, tiling pair (A, B) is generated twice by the in-phase n-bracelet AdB, once by the in-phase n-bracelet $AssB$, once from the out-of-phase bracelet dAB where d covers cells n and 1. If A ends with a square, it is generated by the in-phase n-bracelet AdB; If A ends in a domino, it is generated by the out-of-phase bracelet dAB where d covers cells n and 1. We leave the details for the reader.

Identity 59. Proceed as in the proof of Identity 31. Exploit the fact that $G_2G_0 - G_1^2 = G_3G_0 - G_2G_1$.

Identity 60.

Set 1: Ordered pairs of n-bracelets (A, B). This set has size L_n^2.

Set 2: Ordered pairs of bracelets, (A', B'), where A' has size $n + 1$ and B' has size $n - 1$. This set has size $L_{n+1}L_{n-1}$.

Correspondence: To prove the identity, we establish an almost 1-to-1 correspondence between Set 1 and Set 2. There are exactly five elements that will go unmatched. These five elements are either all in Set 1 or all in Set 2, depending on the parity of n.

We prove this identity when n is even, and leave the odd case for the reader. Let

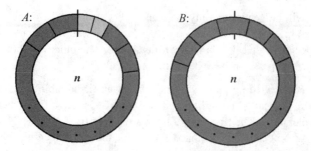

Figure HS.3. The problematic fifth bracelet pair.

A and B be n-bracelets. We exclude from consideration bracelet pairs (A, B) where A and B consist only of dominoes. Since A and B can be of either phase, there are four such bracelet pairs. The fifth bracelet pair we exclude, illustrated in Figure HS.3, is where A begins with two squares, followed by all dominoes and B is the out-of-phase bracelet consisting of all dominoes.

For all other bracelet pairs, there exists a unique number k, $0 \le k < n/2$ such that the last k tiles of A and B are dominoes, but the previous tile of A or B (or both) consists of a square. There are two cases to consider, as illustrated in Figures HS.4 and HS.5.

Case I: B contains a square (which we denote by s) immediately before the last k dominoes. (If $k = 0$, then this says that B ends with a square.) In this case, we transfer s from B to A, where we position s immediately before the last k dominoes of A.

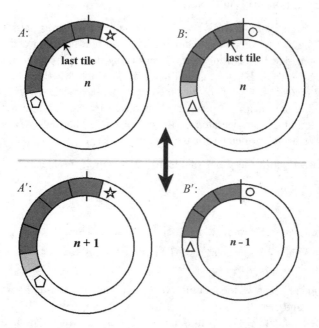

Figure HS.4. Growing and shrinking a pair of n-bracelets. Case I: the last three tiles in A and B are dominoes and the fourth to the last tile in B is a square.

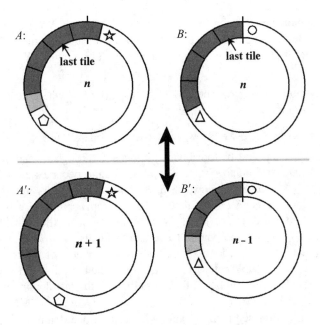

Figure HS.5. Growing and shrinking a pair of n-bracelets. Case II: the last three tiles in A are dominoes, the last four tiles in B are dominoes, and the fourth to the last tile in A is a square.

Case II: B contains a domino (which we denote by d) immediately before the last k dominoes, and A contains a square (denoted by s) immediately before its last k dominoes. In this case, we simply swap tiles d and s.

The transfer procedure of Case I and II results in converting the tiling pair (A, B) to a tiling pair (A', B') where A' has length $n + 1$ and B' has length $n - 1$. Notice that the value of k is the same for (A', B'). Also, notice that A' and B' have the same phase as A and B, respectively. (This would not have been true if we allowed the problematic fifth bracelet among our input. The resulting configuration would have been the same as the one obtained when B was the in-phase all domino configuration.) Hence the procedure is easily reversed. So for n even, $L_n^2 = L_{n+1}L_{n-1} + 5$.

Identity 61. Question: How many phased $2n$-tilings contain at least one square?

Answer 1: There are G_{2n} such tilings minus the G_0 tilings consisting of all dominoes.

Answer 2: Since our board has even length $2n$, the location of the last square must also be even to accommodate the subsequent dominoes. For $1 \le k \le n$, the number of phased tilings whose last square occupies cell $2k$ is G_{2k-1}.

Identity 62. This is the odd length version of the previous identity.

Identity 63. As in the proof of Identity 11, try to break the tiling into $p + 1$ segments where the first p segments have length t if possible and length $t + 1$ if not.

Identity 64. See the proof given for Identity 41.

Identity 65. Let X be an $(n - 1)$-tiling, Y an n-tiling, and W, Z be $(n - 2)$-tilings. For each (X, Y) pair we generate 4 $(n + 1)$-tiling pairs. For each (W, Z) pair we generate one $(n+1)$-tiling pair. Three of the four pairs from (X, Y) are: (Xd, Ys), (Ys, Xd), and (Xss, Ys). The fourth tiling pair from (X, Y) depends on the last tile of Y. Be careful to complete the set of all $(n+1)$-tiling pairs using the last copy of (X, Y) and the pairs (W, Z).

Identity 66. Question: How many phased n-tiling pairs exist where at least one of the tilings contains a domino?

Answer 1: $G_n^2 - G_1^2$, since we throw away the "all-square" tilings.

Answer 2: Let the top and bottom board tile cells 1 through n, and condition on the location of the last domino. Suppose the last domino occupies cells i and $i+1$ for some $1 \leq i \leq n-1$. See Figure HS.6. If this domino occurs in the top board, then we can freely tile cells 1 through $i-1$ of the top board, and cells 1 through $i+1$ of the bottom board, in $G_{i-1}G_{i+1}$ ways. All subsequent tiles must be squares. Otherwise, cell $i+1$ of the top board is a square and cells i and $i+1$ of the bottom board are covered by the last domino. Here, cells 1 through i of the top board may be freely tiled G_i ways, while cells 1 through $i-1$ of the bottom board may be freely tiled G_{i-1} ways. Hence the number of tiling pairs with the last domino beginning at cell i is $G_{i-1}G_{i+1} + G_iG_{i-1} = G_{i-1}(G_{i+1} + G_i) = G_{i-1}G_{i+2}$. Altogether, there are $\sum_{i=1}^{n-1} G_{i-1}G_{i+2}$ phased n-tiling pairs with at least one domino.

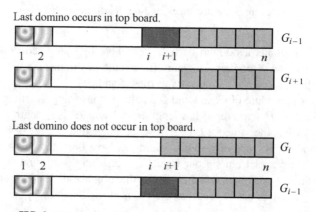

Figure HS.6. The last domino either occurs in the top board or it doesn't.

Identity 67. Question: In how many ways can a phased n-board and a phased $(n-1)$-board be tiled?

Answer 1: G_nG_{n-1}.

Answer 2: We let the top board cover cells 1 through n and the bottom board cover cells 1 through $n-1$. We condition on the last fault, if one exists. First of all, notice that fault-free tilings consist of all dominoes and a single phased square that begins at either the top or bottom board, depending on the parity of n. See Figure HS.7. There are precisely G_0G_1 fault-free tilings. Otherwise, there must exist a fault at some cell i, where $1 \leq i \leq n-1$. For the last fault to occur at cell i, we can freely tile cells 1 through i of both boards G_i^2 ways, followed by the only fault-free way to tile the remaining cells. See Figure HS.8. Altogether our boards may be tiled $G_0G_1 + \sum_{i=1}^{n-1} G_i^2$ ways.

fault-free tilings when n odd

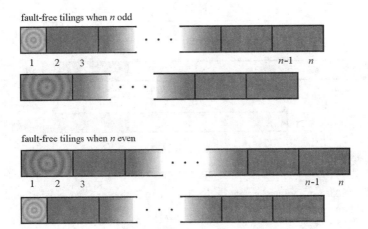

1   2   3 n-1 n

fault-free tilings when n even

1   2   3 n-1 n

Figure HS.7. Two types of fault-free tilings, depending on the parity of n.

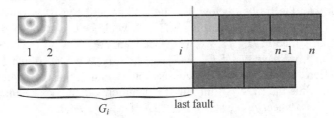

1   2 i n-1 n

G_i last fault

Figure HS.8. There are G_i^2 tilings with last fault at cell i.

Identity 68

Set 1: The set of phased tilings of an m-board on top of an n-board, where the initial conditions of the m board are determined by G_0 and G_1, and the initial conditions of the n-board are determined by H_0 and H_1. This set has size $G_m H_n$.

Set 2: The set of phased tilings of an $(m-1)$-board on top of an $(n+1)$-board, where the initial conditions of the $(m-1)$-board are determined by G_0 and G_1, and the initial conditions of the $(n+1)$-board are determined by H_0 and H_1. This set has size $G_{m-1} H_{n+1}$.

Correspondence: We find an almost one-to-one correspondence between these sets. More precisely, we find a one-to-one correspondence between the "faulty" members of Set 1 and the "faulty" members of Set 2, when they are drawn as in Figure HS.9. After swapping the top tail with the bottom tail, we obtain a phased tiling of Set 2, with the same faults as the original. Since tail swapping is easily reversed (just swap the tiles after the right-most tails again), it follows that

$$G_m H_n - FF1 = G_{m-1} H_{n+1} - FF2,$$

where $FF1$ and $FF2$ denote the number of fault-free tilings in Sets 1 and 2, respectively.

The number of fault-free tilings depends on the parity of m. We first determine $FF1$ when m is even. Here, the m-tiling consists entirely of dominoes. (Note that if

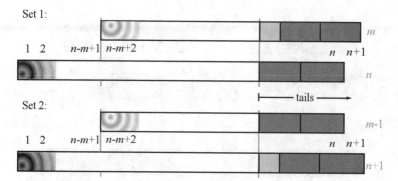

Figure HS.9. After swapping the tails of the first phased tiling pair, we obtain the phased tiling of the second tiling pair.

it began with a square, then it would have to have another square somewhere, creating a fault with the n-tiling below it.) The n-tiling will consist of only dominoes starting with cell $n - m + 3$, see Figure HS.10, while cells 1 through $n - m + 2$ can be arbitrarily tiled in H_{n-m+2} ways. Since there are G_0 ways to choose the all-domino m-tiling, we have that $FF1 = G_0 H_{n-m+2}$. Next we determine $FF2$ when m is even. Here, the $(m - 1)$-tiling consists of an initial phased square followed by dominoes, while the $(n + 1)$-tiling must contain only dominoes beginning at cell $n - m + 2$. Thus, when m is even, $FF2 = G_1 H_{n-m+1}$, yielding

$$G_m H_n - G_0 H_{n-m+2} = G_{m-1} H_{n+1} - G_1 H_{n-m+1},$$

as desired.

Perhaps unexpectedly, when m is odd, $FF1$ and $FF2$ are swapped. That is, as illustrated in Figure HS.11, $FF1 = G_1 H_{n-m+1}$ and $FF2 = G_0 H_{n-m+2}$. Consequently,

$$G_m H_n - G_1 H_{n-m+1} = G_{m-1} H_{n+1} - G_0 H_{n-m+2},$$

as desired.

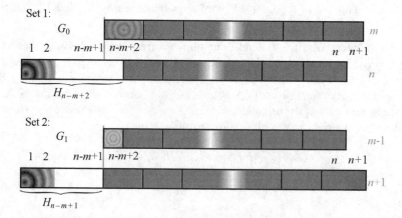

Figure HS.10. Fault-free tilings when m is even.

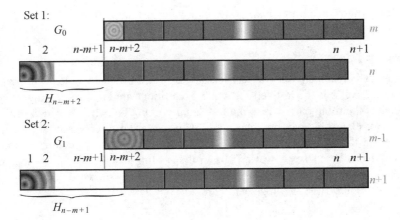

Figure HS.11. Fault-free tilings when m is odd.

Identity 69. See [13].

Identity 70. Every $(n + 1)$-tiling pair and every $(n + 2)$-tiling pair will generate two objects. Let (X, Y) be an $(n + 1)$-tiling pair. It generates (Xd, Yd) and (Xss, Yd). Let (W, Z) be an $(n + 2)$-tiling pair. It generates (Ws, Zs) and another tiling pair depending on the last tiles of W and Z.

For a pretty geometric proof of the Fibonacci version of this identity, see [17].

Exercise 2a. Given an n-bracelet, create the binary sequence $b_1, \ldots, b_n$ where $b_i = 1$ if and only if the tiling is breakable at cell i.

Exercise 2b. Let X be an $(n + 1)$-tiling that does not begin and end with dominoes. If X begins with a square, remove the square to create an in-phase n-bracelet. If X begins with a domino (and ends with a square), remove the final square and create an out-of-phase n-bracelet.

Chapter 3

Identity 72. Follow the proof of Identity 34, but use colored tilings and bracelets. For every n-tiling, we generate $s^2 + 4t$ colored bracelets of length n or $n + 2$. Each colored $(n + 2)$-bracelet is generated once and each n-bracelet is generated t times. Specifically, given a colored n-tiling X, we generate t copies of Bracelet 1. (Applying this to each colored tiling X gives us t copies of every in-phase colored bracelet.) By attaching two colored squares to X in all possible ways (there are s^2 ways to do this), we generate colored versions of Bracelet 2. By attaching all possible colored dominoes to X (there are t ways to do this), we create colored versions of Bracelet 3. Likewise, we generate in t ways, colored versions of Bracelet 4. At this point, we have generated $s^2 + 3t$ colored bracelets. Finally, if X ends in a domino, we create t copies of Bracelet 5a. If X ends with a square we generate, in t ways, colored versions of Bracelet 5b.

Identity 83. Condition on the phase of the colored bracelet.

Identity 84. Condition on the last nonwhite domino.

Identity 85. Condition on the last nonwhite domino.

Identity 86. **Question:** In how many ways can we create a colored n-tiling and a colored $(n+1)$-tiling?

Answer 1: $u_n u_{n+1}$.

Answer 2: For this answer, we ask for $0 \le k \le n$, how many colored tiling pairs exist where cell k is the last cell for which both tilings are breakable? (Equivalently, this counts the tiling pairs where the last square occurs on cell $k+1$ in exactly one tiling.) We claim this can be done $s u_k^2 t^{n-k}$ ways, since to construct such a tiling pair, cells 1 through k of the tiling pair can be tiled u_k^2 ways, the colored square on cell $k+1$ can be chosen s ways, (it is in the n-tiling if and only if $n-k$ is odd) and the remaining $2n - 2k$ cells are covered with $n - k$ colored dominoes in t^{n-k} ways. See Figure HS.12. Altogether, there are

$$s \sum_{k=0}^{n} u_k^2 t^{n-k}$$

tilings, as desired.

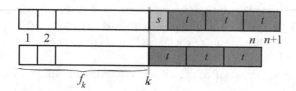

Figure HS.12. A tiling pair where the last mutually breakable cell occurs at cell k.

Identity 87.

Set 1: Tilings of two colored n-boards (a *top* board and a *bottom* board.) By definition, this set has size u_n^2.

Set 2: Tilings of a colored $(n+1)$-board and a colored $(n-1)$-board. This set has size $u_{n+1} u_{n-1}$.

Correspondence: First, suppose n is odd. Then the top and bottom board must each have at least one square. Notice that a square in cell i ensures that a fault must occur at cell i or cell $i - 1$. Swapping the tails of the two n-tilings produces an $(n+1)$-tiling and an $(n-1)$-tiling with the same tails. This produces a 1-to-1 correspondence between all pairs of n-tilings and all tiling pairs of sizes $n + 1$ and $n - 1$ that have faults. Is it possible for a tiling pair with sizes $n + 1$ and $n - 1$ to be fault-free? Yes, with all colored dominoes in staggered formation as in Figure HS.13, which can occur t^n ways. Thus, when n is odd, $u_n^2 = u_{n+1} u_{n-1} - t^n$.

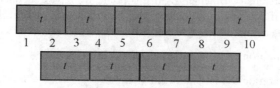

Figure HS.13. When n is odd, the only fault-free tiling pairs consist of all dominoes.

Figure HS.14. When n is even, the only fault-free tiling pairs consist of all dominoes.

Similarly, when n is even, tail swapping creates a 1-to-1 correspondence between faulty tiling pairs. The only fault-free tiling pair is the all domino tiling of Figure HS.14. Hence, $u_n^2 = u_{n+1}u_{n-1} + t^n$. Considering the odd and even case together produces our identity.

Identity 88. Perform tail swapping on a colored n-tiling pair offset by r cells.

Identity 89. Condition on whether the colored $(m + n)$-tiling is breakable at cell m. Use the two copies to account for all bracelet phases in the target bracelet-tiling pairs.

Identity 90. Condition on the phase of the colored $(m + n)$-bracelet and whether it is breakable at cell m. There are lots of cases to consider.

Identity 91. Colorize the proof of Identity 53.

Identity 92. Let every colored n-tiling generate two objects in such a way that every colored $(n-1)$-tiling is generated s times and every colored n-bracelet is generated once. Let X be a colored n-tiling. From X, first generate an in-phase colored n-bracelet. The second bracelet depends on how X begins. If X begins with a domino, then generate an out-of-phase colored n-bracelet. If X begins with a square, remove that square to produce a colored $(n - 1)$-tiling.

Identity 93. Let X be a colored n-tiling. Generate s^2 different colored $(n + 2)$-bracelets that begin with two colored squares. Generate $2t$ colored $(n + 2)$-bracelets by appending an in-phase domino or out-of-phase domino to X. Generate $2t$ more colored $(n + 2)$-bracelets by appending an in-phase domino or out-of-phase domino to X again. Now let Y be a colored $(n + 1)$-bracelet. Use Y to generate s colored $(n + 2)$-bracelets of the sort that have not been generated twice.

Identity 94. Given a colored $(m + n)$-tiling, generate two objects by conditioning on whether the tiling is breakable at cell m and if so, the tile on cell $m + 1$.

Identity 95. Colorize the following construction. Given an $(m + n)$-bracelet B, we create two objects. If B is breakable at cell m, create an m-bracelet, n-bracelet pair and a second bracelet pair or tiling pair, depending on the tiles on cells 1 and $m + 1$. If B is unbreakable at cell m, then generate two tiling pairs (of length $m - 1$ and $n - 1$) which will depend on the phase of B. In this procedure, every tiling pair is created five times. In the colorized version, every tiling pair is created $s^2 + 4t$ times.

Identity 96. Colorize the solution to Identity 53.

Identity 97. **Question:** How many phased colored tilings of two boards of size $2n$ exist, excluding the cases where both boards consist only of dominoes?

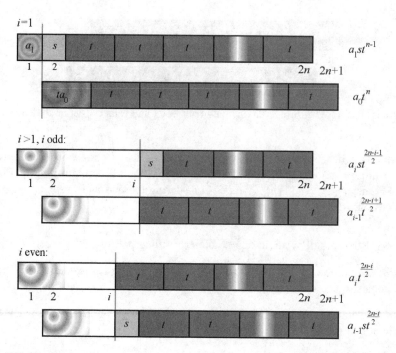

Figure HS.15. A phased colored tiling pairs with last fault at cell i. Three different arguments all lead to the same number.

Answer 1: There are $a_{2n}^2 - t^{2n}a_0^2$ such tilings, since each board can be tiled a_{2n} ways and we throw away the $((ta_0)t^{n-1})^2$ ways where both boards consist only of dominoes.

Answer 2: Let the top board consist of cells 1 through $2n$ and the bottom board consist of cells 2 through $2n + 1$. See Figure HS.15. Since the phased tiling pair has at least one square somewhere, there must be at least one fault that goes through both tilings. Condition on the location of the last fault. We claim that there are $st^{2n-i}a_{i-1}a_i$ ways for the last fault to occur at cell i. The argument differs depending on whether $i = 1$, i is even, or $i > 1$ is odd. When $i = 1$, the top board can be tiled $a_1 st^{n-1}$ ways, and the bottom (all-domino) board can be tiled $(ta_0)t^{n-1}$ ways; hence there are $st^{2n-1}a_1 a_0$ such tilings. When i is even, the top board can be tiled $a_i t^{(2n-i)/2}$ ways, and the bottom board can be tiled $a_{i-1}st^{(2n-i)/2}$ ways, for a product of $st^{2n-i}a_{i-1}a_i$ ways. When $i > 1$ is odd, the top and bottom board can be tiled in $a_i st^{(2n-i-1)/2}$ and $a_{i-1}t^{(2n-i+1)/2}$ ways, respectively, for a product of $st^{2n-i}a_{i-1}a_i$ ways again. Altogether, we have

$$s \sum_{i=1}^{2n} t^{2n-i}a_{i-1}a_i$$

phased colored tiling pairs with at least one square.

Identity 98.

Set 1: The set of phased colored tilings of two $(2n + 1)$-boards. This set has size a_{2n+1}^2.

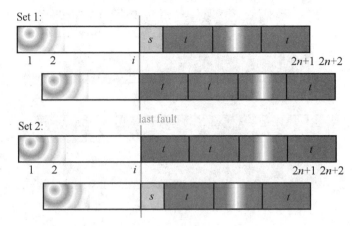

Figure HS.16. After swapping tails of the first phased colored tiling pair, we obtain the phased colored tiling of the second tiling pair.

Set 2: The set of phased colored tiling pairs of a $(2n + 2)$-board and a $(2n)$-board. This set has size $a_{2n+2}a_{2n}$.

Correspondence: We find an almost one-to-one correspondence between the "faulty" members of Set 1 and the "faulty" members of Set 2, when they are drawn as in Figure HS.16. After swapping the top tail with the bottom tail, we obtain a phased tiling of Set 2, with the same faults as the original. Since tail swapping is easily reversed (just swap the tiles after the right-most tails again), it follows that

$$a_{2n+1}^2 - FF1 = a_{2n+2}a_{2n} - FF2,$$

where $FF1$ and $FF2$ denote the number of fault-free tilings in Sets 1 and 2, respectively.

Since $2n + 1$ is odd, any tiling of a $(2n + 1)$-board must contain a square. Thus fault-free tilings in Set 1 happen when the only vertical line to pass through the top and bottom tilings occurs between cells 1 and 2. Since the first tile of the bottom board is phased, no tail swapping is allowed and this concurrence is not a fault. See Figure HS.17. Thus $FF1$ counts the tiling pairs where both boards begin with a phased squared followed by n colored dominoes. So $FF1 = (a_1 t^n)^2$.

Tilings in Set 2 are fault-free when the $(2n)$-board begins with a phased domino followed by all dominoes and the $(2n + 2)$-board begins with a phased domino or a phased square followed by a colored square, and the remaining tiles are dominoes. See Figure HS.18. Thus $FF2 = (ta_0)(t^{n-1})(ta_0 + a_1 s)t^n = a_0^2 t^{2n+1} + a_0 a_1 s t^{2n}$.

Figure HS.17. Fault-free tilings from Set 1.

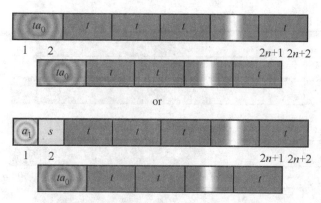

Figure HS.18. Fault-free tilings from Set 2.

Identity 99. Question: How many phased colored $(n+2)$-tilings contain at least one domino?

Answer 1: There are a_{n+2} such tilings minus the $s^{n+1}a_1$ tilings consisting of all squares.

Answer 2: Condition on the location of the last domino. For $0 \leq k \leq n$, there are $a_k t s^{n-k}$ tilings where the last domino covers cells $k+1$ and $k+2$. See Figure HS.19.

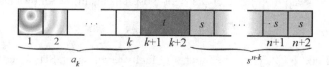

Figure HS.19. Conditioning on the last domino.

Identity 100. Condition on the location of the last square.

Question: How many phased colored $(2n+1)$-tilings exist?

Answer 1: a_{2n+1}.

Answer 2: Since our board has odd length, a last square must exist and occupy an odd cell to accommodate the subsequent dominoes. There are $a_1 t^n$ phased colored tilings where the only square is the initial phased one. Otherwise, for $1 \leq k \leq n$, the number of phased tilings whose last square occupies cell $2k+1$ is $a_{2k} s t^{n-k}$. See Figure HS.20.

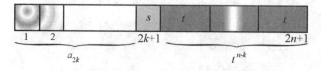

Figure HS.20. The last square of a $(2n+1)$-tiling must occupy an odd cell.

Identity 101. Condition on the location of the last square, if any squares exist.

Identity 102. Condition the last tile that is not a black domino.

Identity 104. a_n counts tilings of an n-board with colored squares, dominoes, and trominoes, where there are c_i choices for a non-initial tile of length i. An initial square has $p_1 = a_1$ phases, an initial domino has $p_2 = a_2 - c_1 a_1$ phases, and an initial tromino has $p_3 = a_3 - c_1 a_2 - c_2 a_1 = c_3 a_0$ phases. The right side counts such tilings of length $2n+2$ that either end with a domino or tromino. Condition on the last non-domino, if any exist. If such a tile begins on cell i, then cells $2i, 2i+1, 2i+2$ are either covered by a tromino (c_3 choices) or a square followed by a domino ($c_1 c_2$ choices).

Identity 105. Similar to previous exercise, but for tilings of odd length.

Other Exercises

Exercise 3.2. For $j > k$, $V_j = c_1 V_{j-1} + c_2 V_{j-2} + \cdots + c_k V_{j-k}$. Choose initial conditions $V_1, V_2, \ldots, V_k$ so that the number of phases for a tile of length i is $p_i = i c_i$. This leads to $V_1 = c_1$ and for $2 \le j \le k$, $V_j = c_1 V_{j-1} + c_2 V_{j-2} + \cdots + c_{j-1} V_1 + c_j j$. (One could also let $V_0 = k$.)

Exercise 3.3. Since u_n has ideal initial conditions, then for $n \ge 0$, u_n counts tilings of n-boards with dominoes and trominoes. The initial conditions for w_n counts tilings with squares and 5-ominoes, but what about the initial tile? Here, $p_1 = w_1 = 1$, $p_2 = w_2 - w_1 = 0$, $p_3 = w_3 - w_2 = 1$, $p_4 = w_4 - w_3 = 0$, and $p_5 = w_0 = 1$. Therefore the initial tile can be a square *or a tromino* or a 5-omino. Let V be a tiling of length n using squares and 5-ominoes. Append two squares to the end of V. Call this new tiling V' which has length $n + 2$. Working from the end of V', we convert V' to a domino-tromino tiling as follows. If V' ends with an even number of squares $2k$, where $k \ge 1$, then convert those squares into k dominoes. (Notation $1^{2k} \to 2^k$.) If V' ends with $2k + 1$ squares, then $1^{2k+1} \to 2^{k-1}3$, meaning that the tiling now ends with a tromino. Moving leftwards, we encounter a 5-omino preceded by m squares, where $m \ge 0$ (notation: $1^m 5$). If m is even, we convert $1^{2k}5 \to 2^{k+1}3$; if m is odd, we convert $1^{2k+1} \to 2^k 3^2$. Continue in the same fashion based on the parity of the number of squares before each 5-omino. When we reach the beginning of the tiling, if it starts with a tromino, then it remains a tromino.

Exercise 3.4. All initial conditions are ideal. Hence g_n counts n-tilings with squares and trominoes, h_n counts n-tilings with dominoes and trominoes, and t_n counts n-tilings with squares, dominoes, and trominoes. (4a) Consider an $(n+3)$-tiling with squares, dominoes, and trominoes. There are f_{n+3} such tilings with no trominoes. Otherwise, condition on the location of the first tromino. Likewise, for (4b) and (4c), condition on the location of the first domino and first square, respectively.

Exercise 3.5. For $1 \le i \le k - 1$, we are restricted to counting a single phased tile. For $i \ge k$, condition on the length of the last tile.

Chapter 4

Exercise 4.2. $\left[a_0, (b_1, a_1), (b_2, a_2), \ldots, (b_n, a_n) \right] = \left[G_1, (G_0, 1), (1, 1), \ldots, (1, 1) \right] = G_{n+1}/f_n$.

Identity 113. Attach a square or unfold a stack of two squares.

Identity 114. If the tiling that satisfies $[a_0, \ldots, a_n, m]$ ends with a stack of m squares on cell $n + 1$, then replace that stack with a domino covering cells $n + 1$ and $n + 2$. Otherwise, append a square on cell $n + 2$.

Identity 115. The denominator is clearly f_n. For the numerator, given an $(n + 1)$-tiling T that can begin with a domino or a stack of up to three squares, we can create an $(n + 2)$-bracelet as follows. If T begins with a domino or a single square, create sT; If T begins with two stacked squares, then "unfold" the first tile to create an in-phase bracelet that begins with a domino; If T begins with three stacked squares, then rotate the previous bracelet one cell to the left to create an out-of-phase bracelet.

Identity 116. For the numerator, modify the argument of the last identity, or just apply reversal to it. The denominator's problem is the same as the numerator's, with one fewer cell.

Identity 117. The denominator counts stackable n-tilings T, where all squares can have up to four squares stacked, except the last one which can have at most three. Basic idea: Given T, each tile gets "tripled" to create a $(3n)$-tiling U in such a way that if the ith cell of T contains one, two or three squares, then U is breakable at cell $3i$; four squares, then U is not breakable at cell $3i$; a domino, then U is breakable at cell $3i$ if and only if the domino ends at cell i. Specifically, suppose for some $k \geq 0$, T begins with k stacks of squares of height 4, followed by a square of height 1 (we denote this by $4^k 1$) then U begins $s^2(ds)^k s$ which is unbreakable at cells $3, 6, 9, \ldots, 3k$, but breakable at cell $3(k + 1)$. Likewise $4^k 2$ generates $d(ds)^k s$; $4^k 3$ generates $s(sd)^k d$ and $4^k d$ generates $d(ds)^k d^2$ which is unbreakable at cells $3, 6, \ldots, 3k + 3$, but breakable at $3k + 6$. Continue through T in this manner.

Identity 118. Same transformation as above, but the final $4^k x$ string in T (where x can be $1, 2, 3, 4$ or d) attaches one extra square at the end (when x is $1, 2, 3$ or d) and $4^k 4$ generates $s^2(ds)^k d$ and $4^k 5$ generates $d(ds)^k d$.

Identity 119. The above fraction is not in lowest terms. The number of n-tilings where all squares can be assigned one of four colors is $f_{3n+2}/2$. To prove this, we need to show that each n-tiling X that satisfies $[4, 4, \ldots, 4]$ generates two uncolored $(3n + 2)$-tilings. If X satisfies $[4, 4, \ldots, 3]$, then we can convert X to a $(3n)$-tiling Y, to which we append a domino or two squares. Otherwise, we need to generate two $(3n + 2)$-tilings that are unbreakable at cell $3n$. To do so, we convert the last $4^k 4$ string to $ss(ds)^k s$ or $d(ds)^k s$.

Identity 120. For the denominator, apply Problem 117. Let T be a tiling with height conditions $[2, 4, \ldots, 4, 3]$. Condition on the first tile of T. T is of the form $1T'$ (a single square followed by T') or $2T'$ (two stacked squares followed by T') or dT'' (A domino followed by T''), where by Problem 117, T' and T'' can be viewed as a traditional $(3n)$-tiling or $(3n - 3)$-tiling, respectively. Now $1T'$ is easily converted to a $(3n + 1)$-bracelet beginning with a square. If T' begins with a square, then $2T'$ becomes an in-phase $(3n+1)$-bracelet beginning with a domino. If T' begins with a domino, then $2T'$ becomes an out-of-phase $(3n + 1)$-bracelet beginning with ds. If T begins with a domino, then dT'' becomes an out-of-phase $(3n + 1)$-bracelet beginning with two dominoes.

Identity 121. Apply the solution strategy of Problem 120 to Problem 118.

Identity 122. Condition on the last tile.

Identity 123. The denominator counts the number of n-tilings with s colored squares and t colored dominoes, which is u_n.

Identity 124. The denominator counts the number of n-bracelets with s colored squares and t colored dominoes, which is v_n.

Chapter 5

Identity 151. Choose a committee of size k and an overseer.

Identity 152. Choose a committee with a president and vice president.

Identity 153. Choose a committee with three distinguished officers.

Identity 154. When picking two distinct ordered pairs, one has either four or three elements involved.

Identity 155. Condition on the number of committees with size m subcommittees.

Identity 156. Condition on the number of even-sized committees with size m subcommittees.

Identity 157. From a class of m men and n women, a size n committee can be created by picking k men to use and k women to *not* use.

Identity 158. Create an upper house and lower house with m distinct members among both houses.

Identity 159. Let $m = n$ in Exercise 5.1 that follows.

Identity 160. When choosing n numbers from 1 to $2n$, the number of odd numbers chosen must equal the number of even numbers unchosen. Why?

Identity 161. If a size n subset contains k "complementary pairs" (that is values of x for which x and $2n + 1 - x$ appear in the subset), then it must contain $n - 2k$ "singletons" and hence there are k complementary pairs of numbers that do not appear in the subset.

Identity 162. Using the binomial coefficient interpretation, condition on largest excluded element. Using the multinomial coefficient interpretation, condition on how many times the element $n + 2$ is used.

Identity 163. Count ordered sets of n phased t-tilings, where the first c of them must begin with a phased square and the remaining $n - t$ of them may begin with a phased square or a phased domino. In the sum, x_i counts the number of dominoes that cover cells i and $i + 1$.

Identity 164. Condition on the median element. For a generalization, condition on the rth smallest element.

Identity 165. A $(2n)$-tiling has at least n tiles and at most $2n$ tiles. If it has $n + k$ tiles, then the number of dominoes is $2n - (n + k) = n - k$ and thus $2k$ squares.

Identity 166. A $(2n - 1)$-tiling has at least n tiles and at most $2n - 1$ tiles. If it has $n + k$ tiles, then the number of dominoes is $2n - 1 - (n + k) = n - k - 1$ and thus $2k + 1$ squares.

Exercise 5.1. Count committees with a leader from the first group.

Exercise 5.2. A path from $(0,0)$ to (a,b) consists of $a + b$ steps. There are $\binom{a+b}{a}$ ways to decide which of those steps will be to the right.

Exercise 5.3a) Condition on the first step.
3b) Each path must pass through exactly one of the points

$$(a,0), (a-1,1), (a-2,2), \ldots, (0,a).$$

3c) Paths through $(s,0), (s-1,1), (s-2,2), \ldots, (0,s)$.
3d) Paths to $(a+1,b)$. Condition on last horizontal step.
3e) Condition on the horizontal step from $x = s$ to $x = s + 1$.
3f) For a solution and history of this identity, see [57].

Exercise 5.4. How many paths that do cross the diagonal? After we reach the first point above the line $(x, x+1)$, "tail swap" so that each horizontal step becomes vertical and each vertical step becomes horizontal, resulting in a path from $(0,0)$ to $(n-1, n+1)$. Hence there are $\binom{2n}{n-1}$ paths that cross the diagonal and therefore

$$\binom{2n}{n} - \binom{2n}{n-1} = \frac{1}{n+1}\binom{2n}{n}$$

paths that do not cross the diagonal. See [52], Chapter 6, Exercise, 19 which gives 66 combinatorially equivalent formulations of the Catalan numbers (and even more are being added to the book's website).

Exercise 5.5. Convert each partition to a path from $(0,0)$ to (a,b).

Exercise 5.6. An ordered partition of n can be viewed as a tiling of length n, where the tiles can have any positive length. Such a tiling is completely determined by which cells are breakable.

Chapter 6

Exercise 5. Consider the set of ordered pairs (S,T) where $S \subseteq \{1, \ldots, n\}$ is a group of students that are explicitly forbidden to be leaders (bad voices perhaps?) and T is the assignment of leaders over the m days where elements are not from S. Let $\mathcal{E}$ and $\mathcal{O}$ denote the set of such (S,T) where the size of S is even and odd, respectively. For a given (S,T), let x denote the largest numbered element of $\{1, \ldots, n\}$ that does not lead. Assuming x exists, we have a bijection $(S,T) \to (S \oplus x, T)$.

Identity 175. Consider the parity of the number of tiles.

Identity 177. How many colored n-bracelets consist of no dominoes where all squares have the same color (black or white)? Let A_i denote the set of colored bracelets with no dominoes where cells i and $i+1$ contain a white square and black square respectively.

Exercise 9. Colored n-bracelets (one color for dominoes and two colors for squares) with an even/odd number of dominoes. Converting the first instance of a domino to a white-black square and vice versa, changes the parity. Two cases are singled out: all black squares or all white squares.

Chapter 7

Identity 195. Distribute n elements into some number of subsets, then place those subsets into m cycles. Excluding the case where $m = n$ and every element is in its own subset and cycle, there must be a smallest such element that is not alone. Use that element to change the parity of the number of subsets.

Identity 198. Condition on the number of elements that do not appear in the same subset as element $n + 1$.

Identity 199. Condition on the largest element that is not alone.

Identity 200. Condition on the largest element that is not alone.

Identity 201. Condition on the smallest element $k + 1$ in the last subset.

Identity 202. Condition on the smallest element $k + 1$ in the last cycle.

Identity 203. The right side counts collections of m ordered lists whose disjoint union is $\{1, \ldots, n\}$. List the lists in increasing order of the first element. The left side counts permutations of $\{1, \ldots, n\}$ where the cycles are placed (in no special order) into m indistinguishable rooms. For definiteness, within a room, arrange the cycles in decreasing order of leading element, then arrange the rooms in increasing order of leading element.

Identity 204. The right side counts the ways n students can be placed into $\ell + m$, classrooms that are indistinguishable, except for the fact that ℓ of them are painted lavender and m of them are painted mauve.

Identity 205. How many ways can you place the numbers $1, 2, \ldots, n$ into ℓ red cycles and m green cycles?

Identity 206. Distribute $n + 1$ elements into at least $m + 1$ subsets. Then except for the subset containing element $n + 1$, place the remaining subsets into m cycles. Proceed as in the previous identity. The excluded case occurs when we have exactly m subsets not containing $n + 1$, and all of these subsets contain just one element.

Identity 207. Combine the proof technique of the Identities 194 and 206.

Identity 208. A subset of students from $\{1, \ldots, n\}$ register for classes. Along with student 0, they are allocated among $m + 1$ indistinguishable (nonempty) classrooms. Let x denote the smallest nonzero element that is either unregistered or in the same classroom as 0, then add or delete x from 0's classroom. The only time this procedure is undefined occurs when 0 is alone and all n students have registered. This occurs $\left\{ {n \atop m} \right\}$ ways.

Identity 209. Students from $\{0, 1, \ldots, n\}$ are to be seated around at least $m + 1$ tables. Then m of the tables (excluding the table with 0) will be selected for prizes. Let x be the smallest positive numbered student that is either at 0's table or at another unwinning table. If x is at 0's table, remove x and all elements to its right to form an unwinning table. Otherwise, merge x's table with 0's table by attaching it to the end of 0's cycle. This procedure is defined, except when 0's table is empty and exactly m other tables have been filled, which occurs $\left[{n \atop m} \right]$ ways.

Identity 210. Among m people who wish to run for city council, some of their names are chosen to appear on the ballot, then an election takes place among n distinguishable voters. Let x denote the largest numbered person among the original m applicants to receive 0 votes.

Note: After re-indexing, and changing the roles of m and n, this can also be done by inclusion-exclusion. See Exercise 1 from Chapter 6. Why does $m!\left\{{n \atop m}\right\}$ count the number of onto functions from $\{1, \ldots, n\}$ to $\{1, \ldots, m\}$?

Challenging Exercise. See [3].

Chapter 8

Identity 222. Count the number of 5-tuples (g, h, i, j, k) where $0 \le g, h, i, j < k \le n$ and condition on how many distinct numbers are used.

Identity 223. Let Set 1 be $\{(i, j, k) | 1 \le i, j < k \le n\}$, and Set 2 be

$$\{(x_1, x_2, x_3) | 1 \le x_1 < x_2 < x_3 \le n', \text{ where } i, j, k \in \{1, 1', 2, 2', \ldots, n, n'\}\},$$

and find a 1-to-4 mapping, based on whether $i < j$ or $i > j$ or $i = j$. For instance, when $i < j$, (i, j, k) maps to $(i, j, k), (i', j, k), (i, j', k), (i', j', k)$.

Identity 224. Part 1) For a length $2n$ binary string with $2k$ 1s to be palindromic (read the same way backwards as forwards), then we must have k 1s in the first half, which can be chosen $\binom{n}{k}$ ways, and the second half is then forced. Each non-palindromic string can be paired up with its reversal. Part 2) A palindromic string must have 1 in the $k+1$st position, and the rest can be chosen $\binom{n}{k}$ ways. Part 3) is similar to part 2. Part 4) All binary strings of length $2n$ with an odd number of 1s must be non-palindromic.

Identity 225. Count the number of length pn binary vectors with pk 1s that are unchanged after all entries are shifted p units to the right.

Identity 226. Same idea as in Identity 225 but don't shift entries $pn+1, pn+2, \ldots, pn+r$.

Identity 227. Consider that $pn+r$ has base p expansion $(a_j, \ldots, a_1, r)_p$ where $(a_j, \ldots, a_1)$ is the base p expansion of n. Begin by choosing pk squares from a rectangle with dimensions p by n. See [51], Chapter 1, problem 6.

Identity 228. Recall L_p counts bracelets of length p. Except for s^p, all tilings have size p orbit.

Identity 229. s^{2p} has an orbit of size one; $\pm d^p$ has an orbit of size two; bracelets of the form xx where x is a p-bracelet have orbits of size p. All other bracelets have orbits of size $2p$.

Identity 230. s^{pq} has an orbit of size one; x^p has an orbit of size q (where x is a q-bracelet not consisting of all squares); x^q has an orbit of size p (where x is a p-bracelet not consisting of all squares). All other bracelets have orbits of size pq.

Theorem 27. In Identity 221, when m divides n, we have $r = 0$ and since $U_0 = 0$, this implies U_m divides U_n.

Theorem 28. *Quick proof.* $L_m f_{m-1} = f_{2m-1}$ divides $f_{2km-1} = F_{2km} = F_n$. *Slower proof.* Count $(2km - 1)$-tiling, conditioning on the first breakable $2jm - 1$ cell. This leads to $F_n = f_{2km-1} = \sum_{j=1}^{k} f_{2m-2}^{j-1} f_{2m-1} f_{2m(k-j)} = F_{2m} \sum_{j=1}^{k} f_{2m-2}^{j-1} f_{2m(k-j)} = L_m f_{m-1} \sum_{j=1}^{k} f_{2m-2}^{j-1} f_{2m(k-j)}$.

Theorem 29. In a length $(2k+1)m$ Lucas tiling, condition on the first breakable cell of the form $(2j+1)m$. This gives us

$$L_{(2k+1)m} = L_m f_{2km} + f_{2m-1} \sum_{j=1}^{k} L_{m-1}(f_{2m-2})^{j-1} f_{2(k-j)m}.$$

Exploit the fact that $f_{2m-1} = f_{m-1}L_m$.

Appendix of Combinatorial Theorems

This is a complete listing of the combinatorial theorems presented throughout the text of this book.

Combinatorial Theorem 1 (p. 1) *Let f_n count the ways to tile a length n board with squares and dominoes. Then f_n is a Fibonacci number. Specifically, for $n \geq -1$,*

$$f_n = F_{n+1}.$$

Combinatorial Theorem 2 (p. 18) *For $n \geq 0$, let ℓ_n count the ways to tile a circular n-board with squares and dominoes. Then ℓ_n is the nth Lucas number; that is*

$$\ell_n = L_n.$$

Combinatorial Theorem 3 (p. 23) *Let $G_0, G_1, G_2, \ldots$ be a Gibonacci sequence with nonnegative integer terms. For $n \geq 1$, G_n counts the number of n-tilings, where the initial tile is assigned a phase. There are G_0 choices for a domino phase and G_1 choices for a square phase.*

Combinatorial Theorem 4 (p. 36) *Let $c_1, c_2, \ldots, c_k$ be nonnegative integers. Let $u_0, u_1, \ldots$ be the sequence of numbers defined by the following recurrence: For $n \geq 1$,*

$$u_n = c_1 u_{n-1} + c_2 u_{n-2} + \cdots + c_k u_{n-k}$$

with "ideal" initial conditions $u_0 = 1$, and for $j < 0$, $u_j = 0$. Then for all $n \geq 0$, u_n counts colored tilings of an n-board, with tiles of length at most k, where for $1 \leq i \leq k$, each tile of length i is assigned one of c_i colors.

Combinatorial Theorem 5 (p. 36) *Let s, t be nonnegative integers. Suppose $u_0 = 1$, $u_1 = s$ and for $n \geq 2$,*

$$u_n = s u_{n-1} + t u_{n-2}.$$

Then for all $n \geq 0$, u_n counts colored tilings of an n-board with squares and dominoes, where there are s colors for squares and t colors for dominoes.

Combinatorial Theorem 6 (p. 36) *Let s, t be nonnegative integers. Suppose $V_0 = 2$, $V_1 = s$ and for $n \geq 2$,*

$$V_n = s V_{n-1} + t V_{n-2}.$$

Then for all $n \geq 0$, V_n counts colored bracelets of length n, where there are s colors for squares and t colors for dominoes.

Combinatorial Theorem 7 (p. 37) *Let s, t, a_0, a_1 be nonnegative integers, and for $n \geq$ 2, define*

$$a_n = sa_{n-1} + ta_{n-2}.$$

For $n \geq 1$, a_n counts the ways to tile an n-board with squares and dominoes where each tile, except the initial one, has a color. There are s colors for squares and t colors for dominoes. The initial tile is given a phase; there are a_1 phases for an initial square and ta_0 phases for an initial domino.

Combinatorial Theorem 8 (p. 37) *Let $c_1, c_2, \ldots, c_k$, $a_0, a_1, \ldots, a_{k-1}$ be nonnegative integers, and for $n \geq k$, define*

$$a_n = c_1 a_{n-1} + c_2 a_{n-2} + \cdots + c_k a_{n-k}.$$

If the initial conditions satisfy

$$a_i \geq \sum_{j=1}^{i-1} c_j a_{i-j}$$

for $1 \leq i \leq k$, then for $n \geq 1$, a_n counts the ways to tile an n-board using colored tiles of length at most k, where each tile, except the initial one, has a color. Specifically, for $1 \leq i \leq k$, each tile of length i may be assigned any of c_i different colors, but an initial tile of length i is assigned one of p_i phases, where

$$p_i = a_i - \sum_{j=1}^{i-1} c_j a_{i-j}.$$

Combinatorial Theorem 9 (p. 51) *Let $a_0, a_1, \ldots$ be a sequence of positive integers, and for $n \geq 0$, suppose the continued fraction $[a_0, a_1, \ldots, a_n]$ is equal to $\frac{p_n}{q_n}$, in lowest terms. Then for $n \geq 0$, p_n counts the ways to tile an $(n+1)$-board with height conditions $a_0, a_1, \ldots, a_n$ and q_n counts the ways to tile an n-board with height conditions $a_1, \ldots, a_n$.*

Combinatorial Theorem 10 (p. 59) *Let $a_0, a_1, \ldots$ be a sequence of positive integers. For $n \geq 1$, suppose the continued fraction $[a_0, (b_1, a_1), \ldots, (b_n, a_n)]$ computed by recurrence (4.9) is equal to $\frac{p_n}{q_n}$. Then for $n \geq 0$, p_n counts the ways to tile an $(n+1)$-board with height conditions $a_0, (b_1, a_1), \ldots, (b_n, a_n)$ and q_n counts the ways to tile an n-board with height conditions $a_1, (b_2, a_2) \ldots, (b_n, a_n)$.*

Combinatorial Theorem 11 (p. 97) *For $n \geq 0$, the nth harmonic number is*

$$H_n = \frac{\left[\frac{n+1}{2}\right]}{n!}.$$

Combinatorial Theorem 12 (p. 141) *The Fibonacci polynomial $f_n(x) = \sum_{k=0}^{n} f(n, k) x^k$, where $f(n, k)$ counts n-tilings with exactly k squares.*

Combinatorial Theorem 13 (p. 142) *The Gibonacci polynomial $G_n(x) = \sum_{k=0}^{n} G(n, k) x^k$, where $G(n, k)$ counts phased n-tilings with exactly k squares, where an initial domino has G_0 phases and an initial square has G_1 phases.*

Appendix of Identities

This is a complete listing of the identities, theorems, and lemmas presented throughout the text of this book.

Identity 1 (p. 2) *For* $n \geq 0$, $f_0 + f_1 + f_2 + \cdots + f_n = f_{n+2} - 1$.

Identity 2 (p. 2) *For* $n \geq 0$, $f_0 + f_2 + f_4 + \cdots + f_{2n} = f_{2n+1}$.

Identity 3 (p. 4) *For* $m, n \geq 0$, $f_{m+n} = f_m f_n + f_{m-1} f_{n-1}$.

Identity 4 (p. 4) *For* $n \geq 0$, $\binom{n}{0} + \binom{n-1}{1} + \binom{n-2}{2} + \cdots = f_n$.

Identity 5 (p. 5) *For* $n \geq 0$, $\sum_{i \geq 0} \sum_{j \geq 0} \binom{n-i}{j} \binom{n-j}{i} = f_{2n+1}$.

Identity 6 (p. 6) *For* $n \geq 0$, $f_{2n-1} = \sum_{k=1}^{n} \binom{n}{k} f_{k-1}$.

Identity 7 (p. 6) *For* $n \geq 1$, $3f_n = f_{n+2} + f_{n-2}$.

Identity 8 (p. 8) *For* $n \geq 0$, $f_n^2 = f_{n+1} f_{n-1} + (-1)^n$.

Identity 9 (p. 8) *For* $n \geq 0$, $\sum_{k=0}^{n} f_k^2 = f_n f_{n+1}$.

Identity 10 (p. 9) *For* $n \geq 0$, $f_n + f_{n-1} + \sum_{k=0}^{n-2} f_k 2^{n-2-k} = 2^n$.

Identity 11 (p. 10) *For* $m, p, t \geq 0$, $f_{m+(t+1)p} = \sum_{i=0}^{p} \binom{p}{i} f_t^i f_{t-1}^{p-i} f_{m+i}$.

Theorem 1 (p. 10) *For* $m \geq 1, n \geq 0$, *if* $m|n$, *then* $F_m | F_n$.

Theorem 2 (p. 11) *For* $m \geq 1, n \geq 0$, *if* m *divides* n, *then* f_{m-1} *divides* f_{n-1}. *In fact, if* $n = qm$, *then* $f_{n-1} = f_{m-1} \sum_{j=1}^{q} f_{m-2}^{j-1} f_{n-jm}$.

Theorem 3 (Euclidean Algorithm) (p. 11) *If* $n = qm + r$, *then* $\gcd(n, m) = \gcd(m, r)$.

Lemma 4 (p. 11) *For* $n \geq 1$, $\gcd(F_n, F_{n-1}) = 1$.

Lemma 5 (p. 12) *For* $m, n \geq 0$, $F_{m+n} = F_{m+1} F_n + F_m F_{n-1}$.

Theorem 6 (p. 12) *For* $m \geq 1, n \geq 0$, $\gcd(F_n, F_m) = F_{\gcd(n,m)}$.

Identity 12 (p. 13) *For* $n \geq 1$, $f_1 + f_3 + \cdots + f_{2n-1} = f_{2n} - 1$.

Identity 13 (p. 13) *For* $n \geq 0$, $f_n^2 + f_{n+1}^2 = f_{2n+2}$.

Identity 14 (p. 13) *For $n \geq 1$, $f_n^2 - f_{n-2}^2 = f_{2n-1}$.*

Identity 15 (p. 13) *For $n \geq 0$, $f_{2n+2} = f_{n+1}f_{n+2} - f_{n-1}f_n$.*

Identity 16 (p. 13) *For $n \geq 2$, $2f_n = f_{n+1} + f_{n-2}$.*

Identity 17 (p. 13) *For $n \geq 2$, $3f_n = f_{n+2} + f_{n-2}$.*

Identity 18 (p. 13) *For $n \geq 2$, $4f_n = f_{n+2} + f_n + f_{n-2}$.*

Identity 19 (p. 13) *For $n \geq 1$, $(f_{n-1}f_{n+2})^2 + (2f_nf_{n+1})^2 = (f_{n+1}f_{n+2} - f_{n-1}f_n)^2 = (f_{2n+2})^2$.*

Identity 20 (p. 13) *For $n \geq p$, $f_{n+p} = \sum_{i=0}^{p} \binom{p}{i} f_{n-i}$.*

Identity 21 (p. 13) *For $n \geq 0$, $\sum_{k=0}^{n}(-1)^k f_k = 1 + (-1)^n f_{n-1}$.*

Identity 22 (p. 13) *For $n \geq 0$, $\prod_{k=1}^{n}\left(1 + \frac{(-1)^{k+1}}{f_k^2}\right) = \frac{f_{n+1}}{f_n}$.*

Identity 23 (p. 13) *For $n \geq 0$, $f_0 + f_3 + f_6 + \cdots + f_{3n} = \frac{1}{2}f_{3n+2}$.*

Identity 24 (p. 13) *For $n \geq 1$, $f_1 + f_4 + f_7 + \cdots + f_{3n-2} = \frac{1}{2}(f_{3n} - 1)$.*

Identity 25 (p. 14) *For $n \geq 1$, $f_2 + f_5 + f_8 + \cdots + f_{3n-1} = \frac{1}{2}(f_{3n+1} - 1)$.*

Identity 26 (p. 14) *For $n \geq 0$, $f_0 + f_4 + f_8 + \cdots + f_{4n} = f_{2n}f_{2n+1}$.*

Identity 27 (p. 14) *For $n \geq 1$, $f_1 + f_5 + f_9 + \cdots + f_{4n-3} = f_{2n-1}^2$.*

Identity 28 (p. 14) *For $n \geq 1$, $f_2 + f_6 + f_{10} + \cdots + f_{4n-2} = f_{2n-1}f_{2n}$.*

Identity 29 (p. 14) *For $n \geq 1$, $f_3 + f_7 + f_{11} + \cdots + f_{4n-1} = f_{2n-1}f_{2n+1}$.*

Identity 30 (p. 14) *For $n \geq 0$, $f_{n+3}^2 + f_n^2 = 2f_{n+1}^2 + 2f_{n+2}^2$.*

Identity 31 (p. 14) *For $n \geq 1$, $f_n^4 = f_{n+2}f_{n+1}f_{n-1}f_{n-2} + 1$.*

Identity 32 (p. 18) *For $n \geq 1$, $L_n = f_n + f_{n-2}$.*

Identity 33 (p. 19) *For $n \geq 0$, $f_{2n-1} = L_n f_{n-1}$.*

Identity 34 (p. 20) *For $n \geq 0$, $5f_n = L_n + L_{n+2}$.*

Identity 35 (p. 21) *For $n \geq 0$, $\sum_{r=0}^{n} f_r L_{n-r} = (n + 2)f_n$.*

Identity 36 (p. 22) *For $n \geq 0$, $L_n^2 = L_{2n} + (-1)^n \cdot 2$.*

Identity 37 (p. 23) *For $n \geq 1$, $G_n = G_0 f_{n-2} + G_1 f_{n-1}$.*

Identity 38 (p. 24) *For $m \geq 1$, $n \geq 0$, $G_{m+n} = G_m f_n + G_{m-1}f_{n-1}$.*

Identity 39 (p. 24) *For $n \geq 0$, $\sum_{k=0}^{n} G_k = G_{n+2} - G_1$.*

Identity 40 (p. 25) *For $n \geq p \geq 0$, $G_{n+p} = \sum_{i=0}^{p} \binom{p}{i} G_{n-i}$.*

Identity 41 (p. 25) *For $n \geq 0$, $\sum_{i=1}^{2n} G_i G_{i-1} = G_{2n}^2 - G_0^2$.*

Identity 42 (p. 26) *For $n \geq 1$, $G_0 H_1 + \sum_{i=1}^{2n-1} G_i H_i = G_{2n} H_{2n-1}$.*

Identity 43 (p. 27) *Let $G_0, G_1, G_2, \ldots$ and $H_0, H_1, H_2, \ldots$ be Gibonacci sequences. Then for $0 \leq m \leq n$, $G_m H_n - G_n H_m = (-1)^m (G_0 H_{n-m} - G_{n-m} H_0)$.*

Identity 44 (p. 28) *Let $G_0, G_1, G_2, \ldots$ and $H_0, H_1, H_2, \ldots$ be Gibonacci sequences. Then for $m, h, k \geq 0$, $G_m H_{m+h+k} - G_{m+h} H_{m+k} = (-1)^m (G_0 H_{h+k} - G_h H_k)$.*

Identity 45 (p. 28) *For $0 \leq m \leq n$, $G_{n+m} + (-1)^m G_{n-m} = G_n L_m$.*

Identity 46 (p. 30) *For $n \geq 1$, $G_{n+1} G_{n-1} - G_n^2 = (-1)^n (G_1^2 - G_0 G_2)$.*

Identity 47 (p. 30) *For $0 \leq m \leq n$, $H_{n-m} = (-1)^m (F_{m+1} H_n - F_m H_{n+1})$.*

Identity 48 (p. 30) *For $n \geq 1$ and $0 \leq m \leq n$,*

$$G_{n+m} - (-1)^m G_{n-m} = F_m (G_{n-1} + G_{n+1}).$$

Identity 49 (p. 31) *For $n \geq 0$, $\sum_{i=0}^{4n+1} G_i = G_{2n+2} L_{2n+1}$.*

Identity 50 (p. 32) *For $n \geq 2$, $L_n = f_{n-1} + 2 f_{n-2}$.*

Identity 51 (p. 32) *For $n \geq 0$, $f_{n-1} + L_n = 2 f_n$.*

Identity 52 (p. 32) *For $n \geq 0$, $5 f_n = L_{n+1} + 2 L_n$.*

Identity 53 (p. 32) *For $n \geq 0$, $5 f_n^2 = L_{n+1}^2 + 4(-1)^n$.*

Identity 54 (p. 32) *For $n \geq 1$, $L_1^2 + L_3^2 + \cdots + L_{2n-1}^2 = f_{4n-1} - 2n$.*

Identity 55 (p. 32) *For $n \geq 0$, $L_{2n+1} - L_{2n-1} + L_{2n-3} - L_{2n-5} + \cdots \pm L_3 \mp L_1 = f_{2n+1}$.*

Identity 56 (p. 32) *For $n \geq 2$, $L_n^4 = L_{n-2} L_{n-1} L_{n+1} L_{n+2} + 25$.*

Identity 57 (p. 32) *For $n \geq 0$, $\sum_{r=0}^{n} L_r L_{n-r} = (n+1) L_n + 2 f_n$.*

Identity 58 (p. 32) *For $n \geq 2$, $5 \sum_{r=0}^{n-2} f_r f_{n-2-r} = n L_n - f_{n-1}$.*

Identity 59 (p. 32) *For $n \geq 2$, $G_n^4 = G_{n+2} G_{n+1} G_{n-1} G_{n-2} + (G_2 G_0 - G_1^2)^2$.*

Identity 60 (p. 32) *For $n \geq 1$, $L_n^2 = L_{n+1} L_{n-1} + (-1)^n \cdot 5$.*

Identity 61 (p. 32) *For $n \geq 0$, $\sum_{k=1}^{n} G_{2k-1} = G_{2n} - G_0$.*

Identity 62 (p. 32) *For $n \geq 0$, $G_1 + \sum_{k=1}^{n} G_{2k} = G_{2n+1}$.*

Identity 63 (p. 32) *For $m, p, t \geq 0$, $G_{m+(t+1)p} = \sum_{i=0}^{p} \binom{p}{i} f_t^i f_{t-1}^{p-i} G_{m+i}$.*

Identity 64 (p. 32) *For $i \geq 1$, $\sum_{i=1}^{n-1} G_{i-1} G_{i+2} = G_n^2 - G_1^2$.*

Identity 65 (p. 32) *For* $n \geq 2$, $G_{n+1}^2 = 4G_{n-1}G_n + G_{n-2}^2$.

Identity 66 (p. 32) *For* $n \geq 1$, $\sum_{i=1}^{n-1} G_{i-1}G_{i+2} = G_n^2 - G_1^2$.

Identity 67 (p. 32) *For* $n \geq 1$, $G_0G_1 + \sum_{i=1}^{n-1} G_i^2 = G_nG_{n-1}$.

Identity 68 (p. 32) *Let* $G_0, G_1, G_2, \ldots$ *and* $H_0, H_1, H_2, \ldots$ *be Gibonacci sequences, then for* $1 \leq m \leq n$, $G_mH_n - G_{m-1}H_{n+1} = (-1)^m[G_0H_{n-m+2} - G_1H_{n-m+1}]$.

Identity 69 (p. 32) $G_{n+1} + G_n + G_{n-1} + 2G_{n-2} + 4G_{n-3} + 8G_{n-4} + \cdots + 2^{n-1}G_0 = 2^n(G_0 + G_1)$.

Identity 70 (p. 32) *For* $n \geq 0$, $G_{n+3}^2 + G_n^2 = 2G_{n+1}^2 + 2G_{n+2}^2$.

Identity 71 (p. 36) *For* $n \geq 1$, $u_{2n-1} = u_{n-1}V_n$.

Identity 72 (p. 37) *For* $n \geq 0$, $(s^2 + 4t)u_n = tV_n + V_{n+2}$.

Identity 73 (p. 38) *For* $m, n \geq 1$, $a_{m+n} = a_mu_n + ta_{m-1}u_{n-1}$.

Identity 74 (p. 39) *For* $n \geq 2$, $a_n - 1 = (s-1)a_{n-1} + (s+t-1)[a_0 + a_1 + \cdots + a_{n-2}]$.

Identity 75 (p. 39) *For any* $1 \leq c \leq s$, *for* $n \geq 0$,

$$a_n - c^n = (s-c)a_{n-1} + ((s-c)c+t)[a_0c^{n-2} + a_1c^{n-3} + \cdots + a_{n-2}].$$

Identity 76 (p. 40) *For* $m, n \geq 2$, $a_{m+n} = a_mu_n + c_2a_{m-1}u_{n-1} + c_3(a_{m-2}u_{n-1} + a_{m-1}u_{n-2})$.

Identity 77 (p. 41) *For* $n \geq 0$, $c_1^n(c_1a_2 + c_3a_0) + (c_1c_2 + c_3)\sum_{i=1}^n c_1^{n-i}a_i = c_1a_{n+2} + c_3a_n$.

Identity 78 (p. 42) *Let* u_n *be defined for* $n \geq 1$ *by* $u_n = u_{n-1} + u_{n-3}$ *where* $u_0 = 1$ *and* $u_j = 0$ *for* $j < 0$. *Then for* $n \geq 0$, $\sum_{i \geq 0} \binom{n-2i}{i} = u_n$.

Identity 79 (p. 42) *Let* u_n *be defined for* $n \geq 1$ *by* $u_n = su_{n-1} + tu_{n-3}$ *where* $u_0 = 1$ *and* $u_j = 0$ *for* $j < 0$. *Then* $\sum_{i \geq 0} \binom{n-2i}{i}t^is^{n-3i} = u_n$.

Identity 80 (p. 42) *Let* u_n *be defined for* $n \geq 1$ *by* $u_n = u_{n-1} + u_{n-3}$ *where* $u_0 = 1$ *and* $u_j = 0$ *for* $j < 0$. *Then*

$$\sum_{a=0}^n \sum_{b=0}^n \sum_{c=0}^n \binom{n-b-c}{a}\binom{n-a-c}{b}\binom{n-a-b}{c} = u_{3n+2}.$$

Identity 81 (p. 43) *Let* g_n *be the* kth *order Fibonacci sequence defined by* $g_j = 0$ *for* $j < 0$, $g_0 = 1$, *and for* $n \geq 1$, $g_n = g_{n-1} + g_{n-2} + \cdots + g_{n-k}$. *Then for all integers* n,

$$g_n = \sum_{n_1} \sum_{n_2} \cdots \sum_{n_k} \frac{(n_1 + n_2 + \cdots + n_k)!}{n_1!n_2!\cdots n_k!},$$

where the summation is over all nonnegative integers $n_1, n_2, \ldots, n_k$ *such that* $n_1 + 2n_2 + \cdots + kn_k = n$.

Identity 82 (p. 44) *Let u_n be the k-bonacci number defined for $n \geq 1$ by $u_n = u_{n-1} + u_{n-k}$, where $u_0 = 1$ and for $j < 0$, $u_j = 0$. Then for $n \geq 0$, $u_{kn+(k-1)}$ equals*

$$\sum_{x_1=0}^{n} \sum_{x_2=0}^{n} \cdots \sum_{x_k=0}^{n} \binom{n-(x_2+x_3+\cdots+x_k)}{x_1} \binom{n-(x_1+x_3+\cdots+x_k)}{x_2} \cdots \binom{n-(x_1+x_2+\cdots+x_{k-1})}{x_k}.$$

Identity 83 (p. 45) *For $n \geq 2$, $V_n = u_n + tu_{n-2}$.*

Identity 84 (p. 45) *For $n \geq 0$,*

$$u_{2n+1} = s[u_0 + u_2 + \cdots + u_{2n}] + (t-1)[u_1 + u_3 + \cdots + u_{2n-1}].$$

Identity 85 (p. 45) *For $n \geq 0$,*

$$u_{2n} - 1 = s[u_1 + u_3 + \cdots + u_{2n-1}] + (t-1)[u_0 + u_2 + \cdots + u_{2n-2}].$$

Identity 86 (p. 46) *For $n \geq 0$, $s\sum_{k=0}^{n} u_k^2 t^{n-k} = u_n u_{n+1}$.*

Identity 87 (p. 46) *For $n \geq 0$, $u_n^2 = u_{n+1}u_{n-1} + (-1)^n t^n$.*

Identity 88 (p. 46) *For $n \geq r \geq 1$, $u_n^2 - u_{n-r}u_{n+r} = (-t)^{n-r+1}u_{r-1}^2$.*

Identity 89 (p. 46) *For $n \geq 1$, $V_n^2 = V_{n+1}V_{n-1} + (s^2 + 4t)(-t)^n$.*

Identity 90 (p. 46) *For $n \geq r \geq 1$, $V_n^2 = V_{n+r}V_{n-r} + (s^2 + 4t)(-t)^{n-r+1}u_{r-1}^2$.*

Identity 91 (p. 46) *For $n \geq 0$, $V_{2n} = V_n^2 - 2(-t)^n$.*

Identity 92 (p. 46) *For $n \geq 1$, $2u_n = su_{n-1} + V_n$.*

Identity 93 (p. 46) *For $n \geq 0$, $(s^2 + 4t)u_n + sV_{n+1} = 2V_{n+2}$.*

Identity 94 (p. 46) *For $m \geq 0$, $n \geq 1$, $2u_{m+n} = u_m V_n + V_{m+1}u_{n-1}$.*

Identity 95 (p. 46) *For $m, n \geq 0$, $2V_{m+n} = V_m V_n + (s^2 + 4t)u_{m-1}u_{n-1}$.*

Identity 96 (p. 46) *For $n \geq 0$, $V_n^2 = (s^2 + 4t)u_{n-1}^2 + 4(-t)^n$.*

Identity 97 (p. 46) *For $n \geq 0$, $a_{2n}^2 - t^{2n}a_0^2 = s\sum_{i=1}^{2n} t^{2n-i}a_{i-1}a_i$.*

Identity 98 (p. 46) *For $n \geq 0$, $a_{2n+1}^2 - a_1^2 t^{2n} = a_{2n+2}a_{2n} - a_0^2 t^{2n+1} - a_0 a_1 s t^{2n}$.*

Identity 99 (p. 46) *For $n \geq 0$, $t\sum_{k=0}^{n} s^{n-k}a_k = a_{n+2} - s^{n+1}a_1$.*

Identity 100 (p. 46) *For $n \geq 0$, $a_{2n+1} = a_1 t^n + s\sum_{k=1}^{n} t^{n-k}a_{2k}$.*

Identity 101 (p. 46) *For $n \geq 0$, $a_{2n} = a_0 t^n + s\sum_{k=1}^{n} t^{n-k}a_{2k-1}$.*

Identity 102 (p. 46) *For $n \geq 1$,*

$$a_{2n+1} = s(a_0 + a_2 + \cdots + a_{2n}) + (t-1)(a_1 + a_3 + \cdots + a_{2n-1}).$$

Identity 103 (p. 46) *For $n \geq 1$,*

$$a_{2n} - 1 = s(a_1 + a_3 + \cdots + a_{2n-1}) + (t - 1)(a_0 + a_2 + \cdots + a_{2n-2}).$$

Identity 104 (p. 46) *For $n \geq 1$,*

$$c_2^n(a_2 - c_1 a_1) + (c_1 c_2 + c_3) \sum_{i=1}^{n} c_2^{n-i} a_{2i-1} = c_2 a_{2n} + c_3 a_{2n-1}.$$

Identity 105 (p. 46) *For $n \geq 1$,*

$$c_2^{n-1}(c_2 a_1 + c_3 a_0) + (c_1 c_2 + c_3) \sum_{i=1}^{n-1} c_2^{n-1-i} a_{2i} = c_2 a_{2n-1} + c_3 a_{2n-2}.$$

Identity 106 (p. 52) *Let $a_0 \geq 0, a_1 > 0, a_2 > 0, \ldots,$ and for $n \geq 0$, let $[a_0, a_1, \ldots, a_n] = p_n/q_n$ in lowest terms. Then*
   a) $p_0 = a_0$, $q_0 = 1$, $p_1 = a_0 a_1 + 1$, $q_1 = a_1$.
   b) For $n \geq 2$, $p_n = a_n p_{n-1} + p_{n-2}$.
   c) For $n \geq 2$, $q_n = a_n q_{n-1} + q_{n-2}$.

Identity 107 (p. 52) *If $a_i = 1$ for all $i \geq 0$, then $[a_0, a_1, \ldots, a_n] = f_{n+1}/f_n$.*

Identity 108 (p. 53) *For all $n \geq 1$, $[2, 1, 1, \ldots, 1, 1, 2] = f_{n+3}/f_{n+1}$, where $a_0 = 2$, $a_n = 2$, and $a_i = 1$ for all $0 < i < n$.*

Identity 109 (p. 53) *Suppose $[a_0, a_1, \ldots, a_{n-1}, a_n] = p_n/q_n$. Then for $n \geq 1$, we have $[a_n, a_{n-1}, \ldots, a_1, a_0] = p_n/p_{n-1}$.*

Identity 110 (p. 54) *The difference between consecutive convergents of $[a_0, a_1, \ldots]$ is: $r_n - r_{n-1} = (-1)^{n-1}/q_n q_{n-1}$. Equivalently, after multiplying both sides by $q_n q_{n-1}$, we have $p_n q_{n-1} - p_{n-1} q_n = (-1)^{n-1}$.*

Identity 111 (p. 55) *$r_n - r_{n-2} = (-1)^n a_n/q_n q_{n-2}$. Equivalently, after multiplying both sides by $q_n q_{n-2}$, we have $p_n q_{n-2} - p_{n-2} q_n = (-1)^n a_n$.*

Identity 112 (p. 57) *For $i < m < j < n$,*

$$K(i, j)K(m, n) - K(i, n)K(m, j) = (-1)^{j-m} K(i, m - 2)K(j + 2, n).$$

Identity 113 (p. 60) *For $n \geq 0$, $[a_0, a_1, \ldots, a_n, 2] = [a_0, a_1, \ldots, a_n, 1, 1]$.*

Identity 114 (p. 60) *For $n \geq 0$, if $m \geq 2$ then*

$$[a_0, a_1, \ldots, a_n, m] = [a_0, a_1, \ldots, a_n, m - 1, 1].$$

Identity 115 (p. 60) *For $n \geq 0$, $[3, 1, 1, \ldots, 1] = L_{n+2}/f_n$, where $a_0 = 3$, and $a_i = 1$ for all $0 < i \leq n$.*

Identity 116 (p. 60) *For $n \geq 1$, $[1, 1, \ldots, 1, 3] = L_{n+2}/L_{n+1}$, where $a_n = 3$, and $a_i = 1$ for all $0 \leq i < n$.*

Identity 117 (p. 60) *For $n \geq 1$, $[4, 4, \ldots, 4, 3] = f_{3n+3}/f_{3n}$, where $a_n = 3$, and $a_i = 4$ for all $0 \leq i < n$.*

Identity 118 (p. 60) *For $n \geq 1$, $[4, 4, \ldots, 4, 5] = f_{3n+4}/f_{3n+1}$, where $a_n = 5$, and $a_i = 4$ for all $0 \leq i < n$.*

Identity 119 (p. 60) *Let $a_i = 4$ for $0 \leq i \leq n$. Then $[4, 4, \ldots, 4] = f_{3n+5}/f_{3n+2}$.*

Identity 120 (p. 60) *For $n \geq 1$, $[2, 4, \ldots, 4, 3] = L_{3n+1}/f_{3n}$, where $a_0 = 2$, $a_n = 3$, and $a_i = 4$ for all $0 < i < n$.*

Identity 121 (p. 60) *For $n \geq 1$, $[2, 4, \ldots, 4, 5] = L_{3n+2}/f_{3n+1}$, where $a_0 = 2$, $a_n = 5$, and $a_i = 4$ for all $0 < i < n$.*

Identity 122 (p. 60) *For nonsimple continued fractions,*

$$P_n = a_n P_{n-1} + b_n P_{n-2},$$
$$Q_n = a_n Q_{n-1} + b_n Q_{n-2}$$

for $n \geq 2$, with initial conditions $P_0 = a_0$, $P_1 = a_1 a_0 + b_1$, $Q_0 = 1$, $Q_1 = a_1$.

Identity 123 (p. 60) *For nonnegative integers s, t, let $u_0 = 1$, $u_1 = s$, and for $n \geq 2$, define $u_n = s u_{n-1} + t u_{n-2}$. Then the nonsimple continued fraction*

$$[a_0, (b_1, a_1), (b_2, a_2), \ldots, (b_n, a_n)] = [s, (t, s), (t, s), \ldots, (t, s)] = u_{n+1}/u_n.$$

Identity 124 (p. 60) *For nonnegative integers s, t, let $v_0 = 2$, $v_1 = s$, and for $n \geq 2$, define $v_n = s v_{n-1} + t v_{n-2}$. Then the nonsimple continued fraction*

$$[a_0, (b_1, a_1), (b_2, a_2), \ldots, (b_{n-1}, a_{n-1}), (b_n, a_n)]$$
$$= [s, (t, s), (t, s), \ldots, (t, s), (2t, s)] = v_{n+1}/v_n.$$

Identity 125 (p. 63) *For $0 \leq k \leq n$, $n! = \binom{n}{k} k! (n-k)!$*

Identity 126 (p. 64) *For $0 \leq k \leq n$, $\binom{n}{k} = \binom{n}{n-k}$.*

Identity 127 (p. 64) *For $0 \leq k \leq n$, (except $n = k = 0$), $\binom{n}{k} = \binom{n-1}{k} + \binom{n-1}{k-1}$.*

Identity 128 (p. 64) *For $n \geq 0$, $\sum_{k \geq 0} \binom{n}{k} = 2^n$.*

Identity 129 (p. 65) *For $n \geq 1$, $\sum_{k \geq 0} \binom{n}{2k} = 2^{n-1}$.*

Identity 130 (p. 65) *For $0 \leq k \leq n$, $k \binom{n}{k} = n \binom{n-1}{k-1}$.*

Identity 131 (p. 66) *For $n \geq 1$, $\sum_{k=0}^{n} k \binom{n}{k} = n 2^{n-1}$.*

Identity 132 (p. 66) *For $m \geq 0$, $n \geq 0$, $\binom{m+n}{k} = \sum_{j=0}^{k} \binom{m}{j} \binom{n}{k-j}$.*

Identity 133 (p. 67) *For $n \geq 0$, $(x+y)^n = \sum_{k=0}^{n} \binom{n}{k} x^k y^{n-k}$.*

Identity 134 (p. 67) *For $0 \leq m \leq k \leq n$, $\binom{n}{k}\binom{k}{m} = \binom{n}{m}\binom{n-m}{k-m}$.*

Corollary 7 (p. 68) *For* $0 < m \leq k < n$, $\binom{n}{m}$ *and* $\binom{n}{k}$ *have a nontrivial common factor. That is,* $\gcd(\binom{n}{m}, \binom{n}{k}) > 1$.

Identity 135 (p. 68) *For* $0 \leq k \leq n$, $\sum_{m=k}^{n} \binom{m}{k} = \binom{n+1}{k+1}$.

Identity 136 (p. 68) *For* $0 \leq k \leq n/2$, $\sum_{m=k}^{n-k} \binom{m}{k}\binom{n-m}{k} = \binom{n+1}{2k+1}$.

Identity 137 (p. 69) *For* $1 \leq r \leq k$, $\sum_{j=r}^{n+r-k} \binom{j-1}{r-1}\binom{n-j}{k-r} = \binom{n}{k}$.

Identity 138 (p. 69) *For* $t \geq 1, n \geq 0$,

$$\sum_{x_1 \geq 0} \sum_{x_2 \geq 0} \cdots \sum_{x_t \geq 0} \binom{n}{x_1}\binom{n-x_1}{x_2}\binom{n-x_2}{x_3}\cdots\binom{n-x_{t-1}}{x_t} = f_{t+1}^n.$$

Identity 139 (p. 70) *For* $t \geq 1, n \geq 0, c \geq 0$,

$$\sum_{x_1 \geq 0} \sum_{x_2 \geq 0} \cdots \sum_{x_t \geq 0} \binom{n-c}{x_1}\binom{n-x_1}{x_2}\binom{n-x_2}{x_3}\cdots\binom{n-x_{t-1}}{x_t} = f_t^c f_{t+1}^{n-c}.$$

Identity 140 (p. 70) *For* $t \geq 1, n \geq 0$,

$$\sum_{x_1 \geq 0} \sum_{x_2 \geq 0} \cdots \sum_{x_t \geq 0} \binom{n}{x_1}\binom{n-x_1}{x_2}\binom{n-x_2}{x_3}\cdots\binom{n-x_{t-1}}{x_t}2^{x_1} = L_{t+1}^n.$$

Identity 141 (p. 70) *For* $t \geq 1, n \geq 0$,

$$\sum_{x_1 \geq 0} \sum_{x_2 \geq 0} \cdots \sum_{x_t \geq 0} \binom{n}{x_1}\binom{n-x_1}{x_2}\binom{n-x_2}{x_3}\cdots\binom{n-x_{t-1}}{x_t}G_0^{x_1}G_1^{n-x_1} = G_{t+1}^n,$$

where G_j *is the jth element of the Gibonacci sequence beginning with* G_0 *and* G_1.

Identity 142 (p. 70) *For* $t \geq 1, n \geq 0$,

$$\sum_{x_1 \geq 0} \sum_{x_2 \geq 0} \cdots \sum_{x_t \geq 0} \binom{n-x_t}{x_1}\binom{n-x_1}{x_2}\binom{n-x_2}{x_3}\cdots\binom{n-x_{t-1}}{x_t} = \frac{f_{tn+t-1}}{f_{t-1}}.$$

Identity 143 (p. 71) *For* $k, n \geq 0$ *and* $k \geq 0$, $\left(\binom{n}{k}\right) = \binom{n+k-1}{k}$.

Identity 144 (p. 72) *For* $0 \leq n \leq m$, $\left(\binom{n}{m-n}\right) = \binom{m-1}{n-1}$.

Identity 145 (p. 72) *For* $n \geq 1, k \geq 0$, $\left(\binom{n}{k}\right) = \left(\binom{k+1}{n-1}\right)$.

Identity 146 (p. 73) *For* $n \geq 0, k \geq 0$ *(except* $n = k = 0$*),* $\left(\binom{n}{k}\right) = \left(\binom{n}{k-1}\right) + \left(\binom{n-1}{k}\right)$.

Identity 147 (p. 73) $k\left(\binom{n}{k}\right) = n\left(\binom{n+1}{k-1}\right)$.

Identity 148 (p. 73) *For* $k \geq 1$, $\left(\binom{n}{k}\right) = \sum_{m=1}^{n} \left(\binom{m}{k-1}\right)$.

Identity 149 (p. 74) *For* $n \geq 0$, $\sum_{k=0}^{m} \left(\binom{n}{k}\right) = \left(\binom{n+1}{m}\right)$.

Identity 150 (p. 74) *For* $n \geq 0$, $\left(\binom{n}{k}\right) = \sum_{m=0}^{n} \binom{n}{m} \left(\binom{m}{k-m}\right)$.

Theorem 8 (p. 75) *For* $n \geq 0$, *the number of odd integers in the nth row of Pascal's triangle is equal to* 2^b *where b is the number of 1s in the binary expansion of* n.

Lemma 9 (p. 75) *Let* r, a, b *be integers where* $r = \frac{a}{b}$. *If a is even and b is odd, then r is even. If a is odd and b is odd, then r is odd.*

Lemma 10 (p. 75) *If n is even and k is odd, then* $\binom{n}{k}$ *is even. Otherwise,*

$$\binom{n}{k} \equiv \binom{\lfloor n/2 \rfloor}{\lfloor k/2 \rfloor} \pmod 2.$$

Identity 151 (p. 78) *For* $n \geq k \geq 0$, $(n-k)\binom{n}{k} = n\binom{n-1}{k}$.

Identity 152 (p. 78) *For* $n \geq 2$, $k(k-1)\binom{n}{k} = n(n-1)\binom{n-2}{k-2}$.

Identity 153 (p. 78) *For* $n \geq 3$, $\sum_{k \geq 0} k(k-1)(k-2)\binom{n}{k} = n(n-1)(n-2)\binom{n-3}{3}$.

Identity 154 (p. 78) *For* $n \geq 4$, $\binom{\binom{n}{2}}{2} = 3\binom{n}{4} + 3\binom{n}{3}$.

Identity 155 (p. 78) *For* $0 \leq m \leq n$, $\sum_{k \geq 0} \binom{n}{k}\binom{k}{m} = \binom{n}{m} 2^{n-m}$.

Identity 156 (p. 78) *For* $0 \leq m < n$, $\sum_{k \geq 0} \binom{n}{2k}\binom{2k}{m} = \binom{n}{m} 2^{n-m-1}$.

Identity 157 (p. 78) *For* $m, n \geq 0$, $\sum_{k \geq 0} \binom{m}{k}\binom{n}{k} = \binom{m+n}{n}$.

Identity 158 (p. 78) *For* $m, n \geq 0$, $\sum_{k \geq 0} \binom{n}{k}\binom{n-k}{m-k} = \binom{n}{m} 2^m$.

Identity 159 (p. 78) *For* $n \geq 1$, $\sum_{k \geq 0} k\binom{n}{k}^2 = n\binom{2n-1}{n-1}$.

Identity 160 (p. 78) *For* $n \geq 0$, $\sum_{k=0}^{n} \binom{n}{k}^2 = \binom{2n}{n}$.

Identity 161 (p. 78) *For* $n \geq 0$, $\sum_{k \geq 0} \binom{n}{2k}\binom{2k}{k} 2^{n-2k} = \binom{2n}{n}$.

Identity 162 (p. 78) *For* $m, n \geq 0$, $\sum_{k=0}^{m} \binom{n+k}{k} = \binom{n+m+1}{m}$.

Identity 163 (p. 78) *For* $t \geq 1, 0 \leq c \leq n$, $(G_1 f_t)^c G_{t+1}^{n-c}$ *equals*

$$\sum_{x_1 \geq 0} \sum_{x_2 \geq 0} \cdots \sum_{x_t \geq 0} \binom{n-c}{x_1}\binom{n-x_1}{x_2}\binom{n-x_2}{x_3}\cdots\binom{n-x_{t-1}}{x_t} G_0^{x_1} G_1^{n-x_1}$$

where G_j *is the jth element of the Gibonacci sequence beginning with* G_0 *and* G_1.

Identity 164 (p. 78) *For* $n, k \geq 0$, $\left(\binom{n}{2k+1}\right) = \sum_{m=1}^{n} \left(\binom{m}{k}\right)\left(\binom{n-m+1}{k}\right)$.

Identity 165 (p. 78) *For* $n \geq 0$, $\sum_{k=0}^{n} \binom{n+k}{2k} = f_{2n}$.

Identity 166 (p. 78) *For* $n \geq 1$, $\sum_{k=0}^{n-1} \binom{n+k}{2k+1} = f_{2n-1}$.

Identity 167 (p. 81) *For $n > 0$, $\sum_{k=0}^{n} \binom{n}{k}(-1)^k = 0$.*

Identity 168 (p. 82) *For $m \geq 0$ and $n > 0$, $\sum_{k=0}^{m} \binom{n}{k}(-1)^k = (-1)^m \binom{n-1}{m}$.*

Theorem 11 (p. 82) *For finite sets $A_1, A_2, \ldots, A_n$, $|A_1 \cup A_2 \cup \cdots \cup A_n|$ is equal to*

$$\sum_{1 \leq i \leq n} |A_i| - \sum_{1 \leq i < j \leq n} |A_i \cap A_j|$$
$$+ \sum_{1 \leq i < j < k \leq n} |A_i \cap A_j \cap A_k| - \cdots + (-1)^n |A_1 \cap A_2 \cap \cdots \cap A_n|.$$

Identity 169 (p. 84) *For $m, n \geq 0$, $\sum_{k=0}^{n} \binom{n}{k}\binom{k}{m}(-1)^k = (-1)^n \delta_{n,m}$.*

Identity 170 (p. 84) *For $n \geq m$, $\sum_{k=0}^{n} \binom{n}{k}\left(\binom{k}{m}\right)(-1)^k = (-1)^n \delta_{n,m}$.*

Identity 171 (p. 85) *For any $m, n \geq 0$, $\sum_{k=0}^{n} \binom{n}{k}\left(\binom{k}{m}\right)(-1)^k = (-1)^n \left(\binom{n}{m-n}\right)$.*

Identity 172 (p. 85) *For $n \geq 0$,*

$$\sum_{k \geq 0} (-1)^k \binom{n-k}{k} = \begin{cases} 1 & \text{if } n \equiv 0 \text{ or } 1 \pmod{6}, \\ 0 & \text{if } n \equiv 2 \text{ or } 5 \pmod{6}, \\ -1 & \text{if } n \equiv 3 \text{ or } 4 \pmod{6}. \end{cases}$$

Identity 173 (p. 86) *For $n \geq 0$, $\sum_{k \geq 0} (-1)^k \binom{n-k}{k} 2^{n-2k} = n + 1$.*

Identity 174 (p. 88) *For $y, m \geq 0$, $\sum_{d=0}^{y} \binom{m+y}{y-d}\left(\binom{m}{d}\right)(-1)^d = 1$.*

Identity 175 (p. 90) *For $n \geq 0$,*

$$\sum_{k \geq 0} (-1)^{n-k} \binom{n-k}{k} = \begin{cases} 1 & \text{if } n \equiv 0 \pmod{3}, \\ 0 & \text{if } n \equiv 2 \pmod{3}, \\ -1 & \text{if } n \equiv 1 \pmod{3}. \end{cases}$$

Identity 176 (p. 90) *For $n \geq 0$,*

$$\sum_{k \geq 0} (-1)^k \frac{n}{n-k} \binom{n-k}{k} = \begin{cases} 2 & \text{if } n \equiv 0 \pmod{6}, \\ 1 & \text{if } n \equiv 1 \text{ or } 5 \pmod{6}, \\ -1 & \text{if } n \equiv 2 \text{ or } 4 \pmod{6}, \\ -2 & \text{if } n \equiv 3 \pmod{6}. \end{cases}$$

Identity 177 (p. 90) *For $n \geq 0$, $\sum_{k \geq 0} (-1)^k \frac{n}{n-k} \binom{n-k}{k} 2^{n-2k} = 2$.*

Identity 178 (p. 91) *For $n \geq 1$, $\sum_{k=1}^{n-1} H_k = n H_n - n$.*

Identity 179 (p. 92) *For $0 \leq m < n$, $\sum_{k=m}^{n-1} \binom{k}{m} H_k = \binom{n}{m+1}\left(H_n - \frac{1}{m+1}\right)$.*

Identity 180 (p. 92) *For $0 \leq m \leq n$, $\sum_{k=m}^{n-1} \binom{k}{m} \frac{1}{n-k} = \binom{n}{m}(H_n - H_m)$.*

Identity 181 (p. 92) *For $n \geq 1$, $\sum_{k=1}^{n-1} k \cdot k! = n! - 1$.*

Identity 182 (p. 93) *For $n \geq 1$, $\sum_{k=1}^{n} \begin{bmatrix} n \\ k \end{bmatrix} = n!$*

Identity 183 (p. 94) *For $n \geq k \geq 1$, $\begin{bmatrix} n \\ k \end{bmatrix} = \begin{bmatrix} n-1 \\ k-1 \end{bmatrix} + (n-1)\begin{bmatrix} n-1 \\ k \end{bmatrix}$.*

Identity 184 (p. 94) *For $n \geq 2$, $\sum_{k=1}^{n} \begin{bmatrix} n \\ k \end{bmatrix}(-1)^k = 0$.*

Identity 185 (p. 95) $x(x+1)(x+2)\cdots(x+n-1) = \sum_{m=1}^{n} \begin{bmatrix} n \\ m \end{bmatrix} x^m$.

Identity 186 (p. 97) *For $n \geq 0$, $\begin{bmatrix} n+1 \\ 2 \end{bmatrix} = n!H_n$.*

Identity 187 (p. 99) *For $n \geq 2$, $\begin{bmatrix} n \\ 2 \end{bmatrix} = (n-1)! + \sum_{k=1}^{n-2} \frac{(n-2)!}{k!} \begin{bmatrix} k+1 \\ 2 \end{bmatrix}$.*

Identity 188 (p. 99) *For $1 \leq t \leq n-1$, $\begin{bmatrix} n \\ 2 \end{bmatrix} = \frac{(n-1)!}{t} + t \sum_{k=t+1}^{n} \begin{bmatrix} k-1 \\ 2 \end{bmatrix} \frac{(n-1-t)!}{(k-1-t)!}$.*

Identity 189 (p. 100) *For $1 \leq m \leq n$, $\begin{bmatrix} n \\ 2 \end{bmatrix} = \begin{bmatrix} m \\ 2 \end{bmatrix} \frac{(n-1)!}{(m-1)!} + \sum_{t=m}^{n-1} \binom{t-1}{m-1} \frac{(m-1)!(n-m)!}{(n-t)}$.*

Theorem 12 (p. 102) *On average, a permutation of n elements has H_n cycles.*

Identity 190 (p. 102) *For $n \geq 1$, $\sum_{k=1}^{n} k\begin{bmatrix} n \\ k \end{bmatrix} = \begin{bmatrix} n+1 \\ 2 \end{bmatrix}$.*

Identity 191 (p. 103) *For $n \geq k \geq 1$, $\begin{Bmatrix} n \\ k \end{Bmatrix} = \begin{Bmatrix} n-1 \\ k-1 \end{Bmatrix} + k\begin{Bmatrix} n-1 \\ k \end{Bmatrix}$.*

Identity 192 (p. 104) *For $n \geq 0$, $x^n = \sum_{k=0}^{n} \begin{Bmatrix} n \\ k \end{Bmatrix} x_{(k)}$.*

Identity 193 (p. 104) *For $k \geq 0$, and for all x sufficiently small,*

$$\sum_{n \geq 0} \begin{Bmatrix} n \\ k \end{Bmatrix} x^n = \frac{x^k}{(1-x)(1-2x)(1-3x)\cdots(1-kx)}.$$

Identity 194 (p. 105) *For $m, n \geq 0$, $\sum_{k=0}^{n} \begin{bmatrix} n \\ k \end{bmatrix}\begin{Bmatrix} k \\ m \end{Bmatrix}(-1)^k = (-1)^n \delta_{m,n}$, where $\delta_{m,n} = 1$ if $m = n$, and is 0 otherwise.*

Identity 195 (p. 106) *For $m, n \geq 0$, $\sum_{k=0}^{n} \begin{Bmatrix} n \\ k \end{Bmatrix}\begin{bmatrix} k \\ m \end{bmatrix}(-1)^k = (-1)^n \delta_{m,n}$.*

Identity 196 (p. 106) *For $m, n \geq 0$, $\sum_{k=m}^{n} \begin{bmatrix} n \\ k \end{bmatrix}\binom{k}{m} = \begin{bmatrix} n+1 \\ m+1 \end{bmatrix}$.*

Identity 197 (p. 106) *For $m, n \geq 0$, $\sum_{k=m}^{n} \begin{bmatrix} n \\ k \end{bmatrix} 2^k = (n+1)!$*

Identity 198 (p. 106) *For $m, n \geq 0$, $\sum_{k=m}^{n} \binom{n}{k}\begin{Bmatrix} k \\ m \end{Bmatrix} = \begin{Bmatrix} n+1 \\ m+1 \end{Bmatrix}$.*

Identity 199 (p. 106) *For $m, n \geq 0$, $\sum_{k=0}^{m} k\begin{Bmatrix} n+k \\ k \end{Bmatrix} = \begin{Bmatrix} m+n+1 \\ m \end{Bmatrix}$.*

Identity 200 (p. 106) *For $m, n \geq 0$, $\sum_{k=0}^{m} (n+k)\begin{bmatrix} n+k \\ k \end{bmatrix} = \begin{bmatrix} m+n+1 \\ m \end{bmatrix}$.*

Identity 201 (p. 106) *For $m, n \geq 0$, $\sum_{k=0}^{n} \begin{Bmatrix} k \\ m \end{Bmatrix}(m+1)^{n-k} = \begin{Bmatrix} n+1 \\ m+1 \end{Bmatrix}$.*

Identity 202 (p. 106) *For $m, n \geq 0$, $\sum_{k=m}^{n} \begin{bmatrix} k \\ m \end{bmatrix} n!/k! = \begin{bmatrix} n+1 \\ m+1 \end{bmatrix}$.*

Identity 203 (p. 106) *For $1 \leq m \leq n$, $\sum_{k=m}^{n} \begin{bmatrix} n \\ k \end{bmatrix}\begin{Bmatrix} k \\ m \end{Bmatrix} = \binom{n}{m}(n-1)!/(m-1)!$*

Identity 204 (p. 106) *For $\ell, m, n \geq 0$, $\sum_{k=0}^{n} \binom{n}{k} \left\{ {k \atop \ell} \right\} \left\{ {n-k \atop m} \right\} = \left\{ {n \atop \ell+m} \right\} \binom{\ell+m}{\ell}$.*

Identity 205 (p. 106) *For $\ell, m, n \geq 0$, $\sum_{k=0}^{n} \binom{n}{k} \left[{k \atop \ell} \right] \left[{n-k \atop m} \right] = \left[{n \atop \ell+m} \right] \binom{\ell+m}{\ell}$.*

Identity 206 (p. 106) *For $m, n \geq 0$, $\sum_{k=m}^{n} \left\{ {n+1 \atop k+1} \right\} \left[{k \atop m} \right] (-1)^k = (-1)^m \binom{n}{m}$.*

Identity 207 (p. 107) *For $0 \leq m \leq n$, $\sum_{k=m}^{n} \left[{n+1 \atop k+1} \right] \left\{ {k \atop m} \right\} (-1)^k = n!/m!$.*

Identity 208 (p. 107) *For $0 \leq m \leq n$, $\sum_{k=m}^{n} \binom{n}{k} \left\{ {k+1 \atop m+1} \right\} (-1)^k = (-1)^n \left\{ {n \atop m} \right\}$.*

Identity 209 (p. 107) *For $0 \leq m \leq n$, $\sum_{k=m}^{n} \left[{n+1 \atop k+1} \right] \binom{k}{m} (-1)^k = (-1)^m \left[{n \atop m} \right]$.*

Identity 210 (p. 107) *For $m, n \geq 0$, $\sum_{k=0}^{m} \binom{m}{k} k^n (-1)^k = (-1)^m m! \left\{ {n \atop m} \right\}$.*

Identity 211 (p. 109) *For $n \geq 0$, $\sum_{k=1}^{n} k = \binom{n+1}{2}$.*

Identity 212 (p. 109) *For $n \geq 0$, $\sum_{k=1}^{n} k = \left(\left({n \atop 2} \right) \right)$.*

Identity 213 (p. 110) *For $n \geq 0$, $\sum_{k=1}^{n} k^3 = \binom{n+1}{2}^2$.*

Identity 214 (p. 111) *For $n \geq 0$, $\sum_{k=1}^{n} k^3 = \left(\left({n \atop 2} \right) \right)^2$.*

Identity 215 (p. 112) *For $n \geq 0$, $\sum_{k=1}^{n} k^2 = \frac{1}{4} \left(\left({2n \atop 3} \right) \right)$.*

Identity 216 (p. 113) *For $n \geq 1$, $(x-1)(1 + x + x^2 + \cdots + x^{n-1}) = x^n - 1$.*

Identity 217 (p. 113) *For $n \geq 0$, $\sum_{k \geq 0} \binom{n}{2k} x^{n-2k} = \frac{1}{2} \left[(x+1)^n + (x-1)^n \right]$.*

Identity 218 (p. 114) *For $n \geq 1$, $\sum_{k=2}^{n} \frac{n!}{k(k-1)} = n! - (n-1)!$*

Theorem 13 (p. 114) *If p is prime, then p divides $\binom{p}{k}$ for $0 < k < p$.*

Lemma 14 (p. 115) *Let g be a function from S to S and let x be an element of S. Suppose for some integer $n \geq 1$, $g^{(n)}(x) = x$ and let m be the smallest positive integer for which $g^{(m)}(x) = x$. Then m divides n.*

Corollary 15 (p. 115) *Let S be a finite set and g be a function from S to S. Suppose n is an integer such that $g^{(n)}(x) = x$ for all x in S. Then the size of every orbit divides n.*

Corollary 16 (p. 115) *Let S be a finite set and suppose there exists a prime number p for which $g^{(p)}(x) = x$ for all x in S. Then every orbit either has size 1 or size p. Consequently, if F is the set of all fixed points of g, then $|S| \equiv |F| \pmod{p}$.*

Theorem 17 (Fermat's Little Theorem) (p. 115) *If p is prime, then for any integer a, p divides $a^p - a$.*

Theorem 18 (p. 116) *If $n = p_1^{e_1} p_2^{e_2} \cdots p_t^{e_t}$, where the p_is are distinct primes and all exponents are positive integers, then $\phi(n) = n \left(1 - \frac{1}{p_1} \right) \left(1 - \frac{1}{p_2} \right) \cdots \left(1 - \frac{1}{p_t} \right)$.*

Corollary 19 (p. 117) *If x and y are integers with no common prime factors, then* $\phi(xy) = \phi(x)\phi(y)$.

Identity 219 (p. 117) $\sum_{d|n} \phi(d) = n$.

Identity 220 (Wilson's Theorem) (p. 117) *If p is prime, then* $(p-1)! \equiv p-1 \pmod{p}$.

Theorem 20 (p. 118) *If G is a group of order n and the prime p divides n, then G has a subgroup of order p.*

Theorem 21 (p. 119) *Let s, t be nonnegative relatively prime integers and consider the sequence $U_0 = 0, U_1 = 1$, and for $n \geq 2$, $U_n = sU_{n-1} + tU_{n-2}$. Then* $\gcd(U_n, U_m) = U_{\gcd(n,m)}$.

Lemma 22 (p. 119) *For all $m \geq 1$, U_m and tU_{m-1} are relatively prime.*

Identity 221 (p. 119) *If $n = qm + r$, where $0 \leq r < m$, then*

$$U_n = (tU_{m-1})^q U_r + U_m \sum_{j=1}^{q} (tU_{m-1})^{j-1} U_{(q-j)m+r+1}.$$

Corollary 23 (p. 120) *If $n = qm + r$, where $0 \leq r < m$, then* $\gcd(U_n, U_m) = \gcd(U_m, U_r)$.

Theorem 24 (p. 120) *Let p be prime. For any $\alpha \geq 1$ and any $0 < k < p^\alpha$, $\binom{p^\alpha}{k} \equiv 0 \pmod{p}$.*

Lemma 25 (p. 121) *For p prime and $\alpha \geq 0$, $(1+x)^{p^\alpha} \equiv 1 + x^{p^\alpha} \pmod{p}$.*

Theorem 26 (Lucas' Theorem) (p. 122) *For any prime p, we can determine $\binom{n}{k} \pmod{p}$ from the base p expansions of n and k. Specifically, if $n = \sum_{i=0}^{t} b_i p^i$ and $k = \sum_{i=0}^{t} c_i p^i$ where $0 \leq b_i, c_i < p$, then $\binom{n}{k} \equiv \prod_{i=0}^{t} \binom{b_i}{c_i} \pmod{p}$.*

Identity 222 (p. 123) *For $n \geq 0$, $\sum_{k=1}^{n} k^4 = \binom{n+1}{2} + 14\binom{n+1}{3} + 36\binom{n+1}{4} + 24\binom{n+1}{5}$.*

Identity 223 (p. 123) *For $n \geq 1$, $\sum_{k=1}^{n-1} k^2 = \frac{1}{4}\binom{2n}{3}$.*

Identity 224 (p. 123) *For $0 \leq r, s \leq 1$ and $n \geq 0$, $\binom{2n+r}{2k+s} \equiv \binom{n}{k}\binom{r}{s} \pmod{2}$.*

Identity 225 (p. 123) *For $n, k \geq 0$ and p prime, $\binom{pn}{pk} \equiv \binom{n}{k} \pmod{p}$.*

Identity 226 (p. 123) *For $0 \leq k \leq n, 0 \leq s \leq r$, and p prime, $\binom{pn+r}{pk+s} \equiv \binom{n}{k}\binom{r}{s} \pmod{p}$.*

Identity 227 (p. 123) *For $0 \leq k \leq n$ and p prime, $\binom{pn}{pk} \equiv \binom{n}{k} \pmod{p^2}$.*

Identity 228 (p. 123) *For p prime, the pth Lucas number satisfies $L_p \equiv 1 \pmod{p}$.*

Identity 229 (p. 123) *For p prime, $L_{2p} \equiv 3 \pmod{p}$.*

Identity 230 (p. 123) *For distinct primes p and q, $L_{pq} \equiv 1 + (L_q - 1)q \pmod{p}$.*

Theorem 27 (p. 123) *If m divides n, then U_m divides U_n.*

Theorem 28 (p. 123) L_m *divides* F_{2km}.

Theorem 29 (p. 123) L_m *divides* $L_{(2k+1)m}$.

Identity 231 (p. 125) *For $n, m \geq 2$,* $f_n f_m - f_{n-2} f_{m-2} = f_{n+m-1}$.

Identity 232 (p. 126) $f_{n-1}^3 + f_n^3 - f_{n-2}^3 = f_{3n-1}$.

Identity 233 (p. 126) $f_n^2 + 2(f_0^2 + f_1^2 + \cdots + f_{n-1}^2) = f_{2n+1}$.

Identity 234 (p. 128) $5[f_0^2 + f_2^2 + \cdots + f_{2n-2}^2] = f_{4n-1} + 2n$.

Identity 235 (p. 130) $\sum_{t=0}^{n} \binom{n}{t} 5^{\lfloor \frac{t}{2} \rfloor} = 2^n f_n$.

Identity 236 (p. 131) $2^{n+1} f_n = \sum_{k=0}^{n} 2^k L_k$.

Identity 237 (p. 132) $2 \sum_{t=0}^{\lfloor \frac{n}{2} \rfloor} \binom{n}{2t} 5^t = 2^n L_n$.

Identity 238 (p. 133) *For $n \geq 1$,* $f_{3n-1} = \sum_{k=1}^{n} \binom{n}{k} 2^k f_{k-1}$.

Identity 239 (p. 135) $G_{3n} = \sum_{j=0}^{n} \binom{n}{j} 2^j G_j$.

Identity 240 (Binet's Formula) (p. 136) *For $n \geq 0$,*

$$F_n = \frac{1}{\sqrt{5}} \left[\left(\frac{1 + \sqrt{5}}{2} \right)^n - \left(\frac{1 - \sqrt{5}}{2} \right)^n \right].$$

Corollary 30 (p. 137) *For $n \geq 0$, f_n is the integer closest to $\phi^{n+1}/\sqrt{5}$.*

Corollary 31 (p. 137) *For $n, m \geq 0$, $\lim_{n \to \infty} \frac{f_{n+m}}{f_n} = \phi^m$.*

Corollary 32 (p. 138) *For $n \geq 1$, $\phi^n = f_n + f_{n-1}/\phi$.*

Corollary 33 (p. 138) *For $n \geq 1$, $\phi^n = \phi f_{n-1} + f_{n-2}$.*

Corollary 34 (p. 138) *For $n \geq 1$, $f_n - \phi f_{n-1} = \frac{(-1)^n}{\phi^n}$.*

Identity 241 (p. 138) *For $n \geq 0$, $L_n = \left(\frac{1+\sqrt{5}}{2} \right)^n + \left(\frac{1-\sqrt{5}}{2} \right)^n$.*

Corollary 35 (p. 139) *For $n \geq 2$, L_n is the integer closest to ϕ^n.*

Identity 242 (p. 139) *For $n \geq 0$, $\phi^n = \frac{\sqrt{5}F_n + L_n}{2}$.*

Identity 243 (p. 139) *For $n \geq 0$, $\left(\frac{-1}{\phi} \right)^n = \frac{L_n - \sqrt{5}F_n}{2}$.*

Identity 244 (p. 139) *For $n \geq 0$, $G_n = \alpha \phi^n + \beta(-1/\phi)^n$, where $\alpha = (G_1 + G_0/\phi)/\sqrt{5}$ and $\beta = (\phi G_0 - G_1)/\sqrt{5}$.*

Identity 245 (p. 142) *For $m, n \geq 1$, $f_{m+n}(x) = f_m(x) f_n(x) + f_{m-1}(x) f_{n-1}(x)$.*

Identity 246 (p. 142) *For $n \geq 1$, $f_n^2(x) - f_{n-1}(x) f_{n+1}(x) = (-1)^n$.*

Bibliography

[1] S.L. Basin and V.E. Hoggatt, Jr., A Primer on the Fibonacci Sequence, Part I, *Fibonacci Quarterly*, **1.1** (1963) 65–72.

[2] Robert Beals, personal correspondence, 1986.

[3] A.T. Benjamin, D.J. Gaebler, and R.P. Gaebler, A Combinatorial Approach to Hyperharmonic Numbers, *INTEGERS: The Electronic Journal of Combinatorial Number Theory*, **3**, #A15, (2003) 1–9.

[4] A.T. Benjamin, C.R.H. Hanusa, F.E. Su, Linear Recurrences through Tilings and Markov Chains, *Utilitas Mathematica*, **64** (2003) 3–17.

[5] A.T. Benjamin, G.M. Levin, K. Mahlburg, and J.J. Quinn, Random Approaches to Fibonacci Identities, *American Math. Monthly*, **107.6** (2000) 511–516.

[6] A.T. Benjamin, G.O. Preston, and J.J. Quinn, A Stirling Encounter with the Harmonic Numbers, *Mathematics Magazine*, **75.2** (2002) 95–103.

[7] A.T. Benjamin, J.D. Neer, D.E. Otero, and J.A. Sellars, A Probabilistic View of Certain Weighted Fibonacci Sums, *Fibonacci Quarterly*, **41.4** (2003) 360–364.

[8] A.T. Benjamin and J.J. Quinn, Recounting Fibonacci and Lucas Identities, *College Math. J.*, **30.5** (1999) 359–366.

[9] A.T. Benjamin and J.J. Quinn, Fibonacci and Lucas Identities through Colored Tilings, *Utilitas Mathematica*, **56** (1999) 137–142.

[10] A.T. Benjamin and J.J. Quinn, The Fibonacci Numbers – Exposed More Discretely, *Mathematics Magazine*, **76.3** (2003) 182–192.

[11] A.T. Benjamin, J.J. Quinn, and J.A. Rouse, "Fibinomial Identities," in *Applications of Fibonacci Numbers*, Vol. 9, Kluwer Academic Publishers, 2003.

[12] A.T. Benjamin, J.J. Quinn, and F.E. Su, Counting on Continued Fractions, *Mathematics Magazine*, **73.2** (2000) 98–104.

[13] A.T. Benjamin, J.J. Quinn, and F.E. Su, Generalized Fibonacci Identities through Phased Tilings, *The Fibonacci Quarterly*, **38.3** (2000) 282–288.

[14] A.T. Benjamin and J.A. Rouse, "Recounting Binomial Fibonacci Identities," in *Applications of Fibonacci Numbers*, Vol. 9, Kluwer Academic Publishers, 2003.

[15] R.C. Brigham, R.M. Caron, P.Z. Chinn, and R.P. Grimaldi, A tiling scheme for the Fibonacci numbers, *J. Recreational Math.*, **28.1** (1996–97) 10–16.

[16] A.Z. Broder, The r-Stirling Numbers, *Discrete Mathematics* **49** (1984) 241–259.

[17] A. Brousseau, Fibonacci Numbers and Geometry, *The Fibonacci Quarterly* **10.3** (1972) 303–318.

[18] P.S. Bruckman, Solution to problem H-487 (proposed by S. Rabinowitz), *The Fibonacci Quarterly* **33.4** (1995) 382.

[19] L. Carlitz, The Characteristic Polynomial of a Certain Matrix of Binomial Coefficients, *The Fibonacci Quarterly* **3.2** (1965) 81–89.

[20] L. Comtet, *Advanced Combinatorics: The Art of Finite and Infinite Expansions*, D. Reidel Publishing Co., Dordrecht, Holland, 1974.

[21] J.H. Conway and R.K. Guy, *The Book of Numbers*, Springer-Verlag, Inc., New York, 1996.

[22] Duane DeTemple, Combinatorial Proofs via Flagpole Arrangements, *College Math. J.*, **35.2** (2004) 129–133.

[23] L.E. Dickson, *History of the Theory of Numbers, Vol. 1*, Carnegie Institution of Washington, 1919.

[24] P. Erdös and R.L. Graham, *Combinatorial Number Theory*, Monographs L'Enseignegnent Mathématique 28, Université de Genève, Geneva, 1980.

[25] A.M. Garsia and S.C. Milne, A Rogers-Ramanujan Bijection, *J. Combinatorial Theory Ser. A* **31** (1981) 289–339.

[26] I.M. Gessel, "Combinatorial Proofs of Congruences," in *Enumeration and Design* ed. D.M. Jackson and S.A. Vanstone, Academic Press, Toronto, 1984.

[27] I.P. Goulden and D.M. Jackson, *Combinatorial Enumeration*, John Wiley & Sons, Inc., New York, 1983.

[28] R.L. Graham, D.E. Knuth, and O. Patashnik, *Concrete Mathematics: A Foundation for Computer Science*, Addison Wesley Professional, New York, 1994.

[29] R.J. Hendel, Approaches to the Formula for the nth Fibonacci Number, *College Math. J.*, **25.2** (1994) 139–142.

[30] D. Kalman and R.A. Mena, The Fibonacci Numbers—Exposed, *Mathematics Magazine*, **76.3** (2003) 167–181.

[31] A. Ya. Khinchin, *Continued Fractions*, University of Chicago Press, 1964.

[32] R. D. Knott, Fibonacci Numbers and the Golden Section,
http://www.mcs.surrey.ac.uk/Personal/R.Knott/Fibonacci/fib.html,
last updated January 17, 2003, last accessed May 2, 2003.

[33] T. Koshy, *Fibonacci and Lucas Numbers with Applications*, John Wiley & Sons, Inc., New York, 2001.

[34] G. Mackiw, A Combinatorial Approach to Sums of Integer Powers, *Mathematics Magazine*, **73** (2000) 44–46.

[35] J.H. McKay, Another Proof of Cauchy's Group Theorem, *American Math. Monthly*, **66.2** (1959) 119.

[36] J.W. Mellor, *Higher Mathematics for Students of Chemistry and Physics*, Dover Publications, New York 1955, p.184.

[37] R.B. Nelsen, *Proof Without Words: Exercises in Visual Thinking*, Classroom Resource Materials, No. 1, MAA, Washington, D.C., 1993.

[38] R.B. Nelsen, *Proof Without Words II: More Exercises in Visual Thinking*, Classroom Resource Materials, MAA, Washington, D.C., 2000.

[39] I. Niven, H.S. Zuckerman, and H.L. Montgomery, *An Introduction to the Theory of Numbers*, John Wiley & Sons, Inc., New York, 1991.

[40] O. Perron, *Die Lehre von den Kettenbrüchen*, Chelsea Publishing Co., 1929.

[41] G. Pólya, R.E. Tarjan, and D.R. Woods, *Notes on Introductory Combinatorics*, Birkhauser, Boston, 1983.

[42] C. Pomerance and A. Sárközy, "Combinatorial Number Theory," in *Handbook of Combinatorics* Vol. 1, Elsevier, Amsterdam, 1995, 967–1018.

[43] Greg Preston, *A Combinatorial Approach to Harmonic Numbers*, Senior Thesis, Harvey Mudd College, Claremont, CA, 2001.

[44] H. Prodinger and R.F. Tichy, Fibonacci Numbers of Graphs, *Fibonacci Quarterly*, **20** (1982) 16–21.

[45] J. Propp, A Reciprocity Theorem for Domino Tilings, *Electron. J. Combin.* **8.1** (2001) R18, 9 pp.

[46] J. Riordan, *Combinatorial Identities*, John Wiley & Sons, Inc., New York, 1968.

[47] G.-C. Rota and B.E. Sagan, Congruences Derived from Group Action, *European J. Combin.* **1** (1980) 67–76.

[48] J. Rouse, *Combinatorial Proof of Congruences*, Senior Thesis, Harvey Mudd College, Claremont, CA, 2003.

[49] B.E. Sagan, Congruences via Abelian Groups, *J. Number Theory* **20.2** (1985) 210–237.

[50] J.H. Smith, Combinatorial Congruences from p-subgroups of the Symmetric Group, *Graphs Combin.* **9.3** (1993) 293–304.

[51] R. Stanley, *Enumerative Combinatorics Vol. 1*, Cambridge University Press, 1997.

[52] R. Stanley, *Enumerative Combinatorics Vol. 2*, Cambridge University Press, 1999.

[53] D. Stanton and D. White, *Constructive Combinatorics*, Springer-Verlag, Inc., New York, 1986.

[54] W. Staton and C. Wingard, Independent Sets and the Golden Ratio, *College Math. J.*, **26.4** (1995) 292–296.

[55] M. Sved, Tales from the "County Club", *Lecture Notes in Mathematics # 829, Combinatorial Mathematics VII*, Springer-Verlag, Inc., New York, 1980.

[56] M. Sved, Counting and Recounting, *Mathematical Intelligencer* **5** (1983) 21–26.

[57] M. Sved, Counting and Recounting: The Aftermath, *The Mathematical Intelligencer*, **6** (1984) 44–45.

[58] S. Vajda, *Fibonacci & Lucas Numbers, and the Golden Section: Theory and Applications*, John Wiley & Sons, Inc., New York, 1989.

[59] M.E. Waddill, "Using Matrix Techniques to Establish Properties of a Generalized Tribonacci Sequence," in *Applications of Fibonacci Numbers*, Vol. 4, Kluwer Academic Publishers, 1991, 299–308.

[60] M. Werman and D. Zeilberger, A Bijective Proof of Cassini's Fibonacci Identity, *Discrete Math.*, **58** (1986) 109.

[61] H. Wilf, *generatingfunctionology*, Elsevier Academic Press, Amsterdam, 1994.

[62] R.M. Young, *Excursions in Calculus: An Interplay of the Continuous and the Discrete*, Dolciani Math. Exp. 13., MAA, Washington, D.C., 1992, Chapter 3.

[63] D. Zeilberger, Garsia and Milne's Bijective Proof of the Inclusion-Exclusion Principle, *Discrete Math.*, **51** (1984) 109–110.

Index

About the Authors

Arthur T. Benjamin was born in Cleveland, Ohio. He received his BS in Applied Mathematics from Carnegie Mellon. In 1989, he received his PhD in Mathematical Sciences from Johns Hopkins University, and was awarded the Nicholson Prize from the Operations Research Society of America. Currently, Art is Professor and Chair of the Mathematics Department at Harvey Mudd College, and is a Fellow of the Institute for Combinatorics and its Applications. In 2000, he was awarded the Haimo Prize from MAA for Distinguished Teaching.

Art Benjamin is also a professional magician and lightning calculator, frequently performing at The Magic Castle in Hollywood and for uncountably many schools and organizations around the world. He is an Associate Editor of *Mathematics Magazine, The UMAP Journal,* and past editor of the Spectrum book series for MAA. He and Jennifer Quinn are co-editors of *Math Horizons.* Art lives in Claremont, California with his wife Deena and daughters Laurel and Ariel.

Jennifer J. Quinn was born in Princeton, NJ though she spent the bulk of her formative years growing up in Rhode Island. She received her BA magna cum laude from Williams College with majors in Mathematics and Biology and was elected to the Phi Beta Kappa Society. Working as an actuary in Chicago after graduation convinced her to return to academic life. In 1987, she earned her MS in Pure Mathematics from the University of Illinois at Chicago and in 1993, earned her PhD in Combinatorics from the University of Wisconsin, Madison. She has been working at Occidental College ever since where currently she is an Associate Professor and Chair of the Mathematics Department.

Jennifer Quinn has been a visiting member of the Mathematical Sciences Research Institute and the Institute for Mathematics and its Applications. She was awarded an NSF Course and Curriculum Development Grant to create an interdisciplinary introductory calculus and mechanics course, received the 2001 Distinguished Teaching Award from the Mathematical Association of America for the Southern California Section, and served as President of California's Delta Chapter of Phi Beta Kappa. She is a Fellow of the Institute of Combinatorics and its Applications. Regionally, she has served on the Board of the Southern California Section of the MAA as Program Chair and Newsletter Editor, and the Phi Beta Kappa Alpha Alumni Association of California as Councilor, Chair of the International Scholarship Committee, and Vice President of the Board. Nationally, she is an Associate Editor for *Mathematics Magazine,* past editorial board member for the Spectrum book series of the MAA, and co-editor of *Math Horizons.* She lives in Los Angeles, California with her husband Mark Martin, a microbial geneticist—not a racecar driver, and sons Anson and Zachary.